John Seymour St. John

Larva Collecting and Breeding

A Handbook to the Larvae of the British Macro-Lepidoptera and their Food Plants

John Seymour St. John

Larva Collecting and Breeding
A Handbook to the Larvae of the British Macro-Lepidoptera and their Food Plants

ISBN/EAN: 9783337144029

Printed in Europe, USA, Canada, Australia, Japan

Cover: Foto ©berggeist007 / pixelio.de

More available books at **www.hansebooks.com**

WILLIAM WESLEY & SON,

28, ESSEX STREET, STRAND, LONDON.
PUBLISHERS AND BOOKSELLERS.

Spécialité :—Works on **Natural History** and **Science**.
Agents for **The Smithsonian Institution**.

THE FOLLOWING RECENTLY-PUBLISHED

Natural History & Scientific Book Circulars

INCLUDE A PORTION OF THEIR STOCK :—

No. 95. Geology and Mineralogy; including the first part of the Library of the late W. H. BAILY, of the Geological Survey of Ireland. 2500 Works. Price 6*d*.

No. 96. Palæontology; including the second part of the Library of the late W. H. BAILY. Palæontologist to the Geological Survey of Ireland. About 1000 Works. Price 4*d*.

No. 97.—Ichthyology; Reptilia and Amphibia; General Zoology, Darwinism, Anatomy, Physiology, and Embryology. Price 4*d*.

No. 98.—Ornithology; Local and British Ornithological Faunas; Foreign and Exotic Ornithological Faunas; Cage Birds and Poultry; Ornithological Monographs; **Mammalia; Faunas and Geography; Zoological Voyages.** Over 1000 Works. Price 4*d*.

No 99.—Astronomy, Mathematics, Physics. Price 2*d*.

No. 100.—Entomology :—Lepidoptera, Coleoptera, Hymenoptera, Diptera, Neuroptera, Orthoptera, Hemiptera. Arachnida, Aptera (Parasitel), Injurious and Useful Insects, Systematic and General Entomology. 1000 Titles.

Orders for Works on **Natural History** *and* **Science**, *whether in the Book Circular or not, will be carefully attended to:* W. Wesley & Son *have great facilities for obtaining works of this class.*

RECENT PUBLICATIONS.

Comstock (J. H.)—Introduction to Entomology; Part I. (to be completel in two parts), 201 original engravings, 8vo., 10s.

Gurney, Russell and Coues.—The House Sparrow, plate, crown 8vo., cloth, price, 2s. 6*d*.

"The information just published by Mr. Gurney and Col. Russell will be welcomed by all who desire to understand the merits of the question, and to arrive at a right conclusion."—*From* "*The Field.*"

Just Published, Parts 1 and 2, with 6 Coloured Plates, 4to, 5s. each,
Post Free.

NORTH AMERICAN BIRDS. By H. NEHRLING. To be completed in 12 Parts, containing 36 Coloured Plates after Water-colour Drawings by Prof. Robert Ridgway, of the United States National Museum, A. Goering, and Gustav Muetzel.

W. WESLEY & SON, 28, Essex Street, Strand, London.

THE PRACTICAL CABINET MAKER.

A PRICE LIST
OF
CABINETS AND APPARATUS
OF EVERY DESCRIPTION FOR THE USE OF
ENTOMOLOGISTS, ORNITHOLOGISTS, BOTANISTS, &c.

BIRDS' EGGS AND SKINS,
British and Foreign Butterflies and Moths.

Apparatus and Cabinets of every description of the best make for Entomology and general Natural History.
Ring Net (Wire or Cane) and Stick, 1/8, 2/-, 2/3.
Umbrella Net (self-acting), 7/6; Pocket Folding Net (Wire or Cane), 3/9, 4/6.
Corked Pocket Boxes, 6d., 9d. 1/-, 1/6; Zinc Relaxing Boxes, 1/-, 1/6, 2/-.
Chip Boxes (nested), 4 doz., 8d. Entomological Pins, mixed, 1/- per oz.
Pocket Lantern, 2/6, 5/-, 10/6. Sugaring Tin (with brush), 1/6, 2/-.
Mite Destroyer, 1/6 per lb. Killing Bottles, 1/6.
Store Boxes, 2/6, 4/-, 5/-, 6/-. Setting Boards from 6d.
Setting Houses, 9/6, 11/6, 14/-. Larva Boxes, 9d., 1/-, 1/-.
Breeding Cages, 2/6, 4/-, 5/-, 7/6.
A very large stock of British and Foreign Lepidoptera, Coleoptera, Birds' Eggs, etc.
Works on Entomology, New and Second-hand.
A complete Catalogue sent on application, post free.

J. T. CROCKETT,
Special Show Rooms,
7a, Princes Street, Cavendish Square,
LONDON, W.
(Seven doors from Oxford Circus.)

Factories—34, Ridinghouse St., and Ogle St., W.

LARVA COLLECTING

AND

BREEDING:

A HANDBOOK TO THE LARVÆ

OF THE

BRITISH MACRO-LEPIDOPTERA

AND THEIR FOOD PLANTS;

BOTH IN NATURE AND IN CONFINEMENT,

WITH AUTHORITIES.

BY THE

REV. J. SEYMOUR ST. JOHN, B.A.

LONDON
WILLIAM WESLEY AND SON,
28, ESSEX STREET, STRAND.

AND TO BE HAD OF THE AUTHOR,
42, CASTLEWOOD ROAD, STAMFORD HILL, N.
1890.

HAYMAN, CHRISTY AND LILLY, LTD.,
PRINTERS,
HATTON WORKS, 113, FARRINGDON ROAD,
AND 20, 22, ST. BRIDE ST., E.C.

PREFACE.

IN introducing this book to the entomological world, my object is to render some help to those who make Lepidoptera—especially in their second or larval stage—their pleasant and interesting study; and also to place in a clear and concise form, easy for reference, such information concerning the food-plants of the larvæ of our British Macrolepidoptera, as I have been able to gather from various published works, competent friends, and my own personal experience.

The compilation of this work is no claim to great personal knowledge and very wide experience of the habits and food-plants of the larvæ in question, but has been rather attained by patient care, considerable labour, and the kind assistance of a few entomologists. I do not profess to have given *all* the food-plants upon which the larvæ have been recorded or are said to feed; for (i.) the *natural* food-plants of certain very rare or local larvæ are as yet either unknown or still remain a secret to the few; (ii.) some plants or trees have been recorded as the "pabulum in naturâ" of certain larvæ found upon them at rest or "nibbling" at them through lack of their proper food; while (iii.) in the case of polyphagous or semi-polyphagous larvæ, two or three food-plants are deemed sufficient to name, with a note of their being general feeders, in order not to add unnecessarily to the bulk of the book.

The book, being intended primarily as a *book of reference*, is published in a form most convenient for carrying about on collecting expeditions.

The larvæ in the first portion of the book are arranged and named according to "The Entomologist Synonymic List of British Lepidoptera," with one or two recently-discovered species inserted; the names (generic and specific) given in brackets are those of the "Doubleday List" nomenclature.

The arrangement and nomenclature of the food-plants in the second portion of the book are according to "The London

Catalogue of British Plants" (8th ed.). The numbers placed after the names of the plants, etc., are taken from the same Catalogue, and are (as explained in it) intended for a *census*, *i.e*, "a scale of rarity or frequency in relation to Britain as a whole. They express the number of counties in which each species has been reputed to occur, as set forth in the 2nd edition of *Topographical Botany*. By sub-dividing the larger counties into two or more vice-counties" their number amounts to 112. So that, for example, the number "96" placed after "*Malva sylvestris*.—Mallow," means that that plant grows in 96 out of the 112 districts into which Britain is divided.

References have been made with regard to a few species, especially some of the rarer larvæ, and those of "occasional visitors" to Britain (marked with an asterisk), to W. F. Kirby's "European Butterflies and Moths," Kaltenbach's "Pflanzenfeinde," and to Owen S. Wilson's "Larvæ of the British Lepidoptera and their Food-plants."

A few blank pages are added to the book for private notes and memoranda.

The following abbreviations are used:—

Buckl.—Larvæ of British Butterflies and Moths.—Buckler and Stainton.

Newm.—Natural History of British Butterflies and Moths. —E. Newman.

Sta.—Manual of British Butterflies and Moths.—H. T. Stainton.

Ent.—"The Entomologist."

Ent. Mag.—"The Entomologist's Monthly Magazine."

In conclusion, I tender, first of all, my warmest and most grateful thanks to my friend, Mr. F. J. Hanbury, for his unvarying kindness and valuable assistance in many ways, especially as regards the botanical matter; and secondly, my sincere thanks for information so readily afforded me, to Mrs. Hutchinson and Messrs. E. A. Atmore, W. H. Harwood, W. R. Jeffrey, W. Machin, F. Norgate, Richard South, and Sydney Webb.

J. SEYMOUR ST. JOHN.

42, CASTLEWOOD ROAD,
    STAMFORD HILL, N.
        March, 1890.

LARVA COLLECTING AND BREEDING.

LARVÆ AND FOOD-PLANTS.

RHOPALOCERA.

Papilio machaon.
    Peucedanum palustre—Milk Parsley.
    Heracleum Sphondylium—Cow-parsnep.
    Angelica sylvestris—Wild Angelica.
    In confinement on carrot-leaves and rue.

Aporia cratægi.
    Prunus—Plum.
    Pyrus Malus—Apple.
       „ *communis*—Pear.
    Cratægus Oxyacantha—Whitethorn.

Pieris brassicæ.
    Tropæolum majus.
    Brassica Napus—Rape.
    Also on various cultivated cabbage.

Pieris rapæ.
    Nasturtium officinale—Water-cress.
    Sisymbrium officinale—Hedge Mustard.
    Brassica Napus—Rape.
       „ *Sinapis*—Wild Mustard.
    Reseda lutea—Wild Mignonette.
    Tropæolum majus.
       „ *canariense.*
    Also on cabbage, mignonette and horse-radish.

Pieris napi.
    Nasturtium officinale—Water-cress.
    Barbarea vulgaris—Winter Cress.
       „   *præcox*—Early Winter Cress.
    Cardamine pratensis—Cuckoo-flower.
    Sisymbrium Alliaria—Jack-by-the-hedge.
    Brassica Napus—Rape.
    Cakile maritima—Sea Rocket.
    Also on horse-radish and mignonette.

Pieris daplidice.
    Reseda Luteola—Dyer's Rocket.
      „   *lutea*—Wild Mignonette.
    In confinement on *Reseda odorata*.

Euchloë (Anthocharis) cardamines.
    Nasturtium officinale—Water-cress.
    Barbarea vulgaris—Winter Cress. On the seeds.
    Arabis perfoliata—Rock Cress.
    Cardamine pratensis—Cuckoo-flower.
      „   *impatiens*—Narrow-leaved Bitter Cress.
    Hesperis matronalis—Dame's Violet.
    Sisymbrium officinale—Hedge Mustard.
      „   *Alliaria*—Jack-by-the-hedge.
    Brassica Sinapis—Wild Mustard.
    Also on the flower-stems of horse-radish.

Leucophasia sinapis.
    Lotus corniculatus—Bird's-foot Trefoil.
      „   *pilosus*—Greater Bird's-foot Trefoil.
    Vicia Cracca—Tufted Vetch.
    Lathyrus (Orobus) tuberosus—Tuberous Bitter Vetch.
      „   „   *sylvestris*—Everlasting Pea.

Colias hyale.
    Medicago sativa—Lucerne.
      „   *lupulina*—Black Medick.
    Trifolium pratense—Purple Clover.
      „   *repens*—Dutch Clover.
    Lotus corniculatus—Bird's-foot Trefoil.
    Also on other leguminous plants.

Colias edusa.
    Trifolium pratense—Purple Clover.
      „   *repens*—Dutch Clover.
    Lotus corniculatus—Bird's-foot Trefoil.
    Also on other leguminous plants.

Gonopteryx rhamni.
    Rhamnus Catharticus—Buckthorn.
      „   *Frangula*—Alder Buckthorn.

LARVÆ AND FOOD-PLANTS.

Argynnis selene.
   Viola canina—Dog Violet.

Argynnis euphrosyne.
   Viola canina—Dog Violet.
   Primula acaulis—Primrose. In confinement, after hybernation.

Argynnis latona.
   Viola odorata—Sweet Violet.
   „ *canina*—Dog Violet.
   „ *tricolor*—Heartsease.

Argynnis aglaia.
   Viola canina—Dog Violet.

Argynnis adippe.
   Viola canina—Dog Violet.
   „ *tricolor*—Heartsease.

Argynnis paphia.
   Viola odorata—Sweet Violet.
   „ *canina*—Dog Violet.

* **Argynnis dia.**
   Viola canina—Dog Violet.
   „ *tricolor*—Heartsease.

* **Argynnis niobe.**
   Viola canina—Dog Violet.
   „ *tricolor*—Heartsease.

Melitæa aurinia (artemis).
   Lonicera Periclymenum—Honeysuckle.
   Scabiosa succisa—Devil's-bit Scabious.
   Plantago—Plantain.

Melitæa cinxia.
   Plantago maritima—Sea Plantain.
   „ *Coronopus*—Buck's-horn Plantain.
   Also on species of Hieracium and Veronica.

Melitæa athalia.
   Digitalis purpurea—Foxglove.
   Veronica Chamædrys—Germander Speedwell.
   Melampyrum pratense—Cow-wheat.
   „ *sylvaticum*—Wood Cow-wheat.
   Teucrium Scorodonia—Wood Sage.
   Plantago major—Greater Plantain.
   „ *lanceolata*—Narrow-leaved Plantain.

* **Melitæa didyma.**
   Plantago—Plantain.

Vanessa c-album.
Ribes rubrum—Red Currant.
Ulmus campestris—Elm.
Humulus Lupulus—Hop.
Urtica dioica—Stinging Nettle.

Vanessa polychloros.
Prunus Avium—Wild Cherry.
Pyrus Aria—White-Beam.
„ *communis*—Pear.
Ulmus montana—Wych Elm.
„ *campestris*—Elm.
Salix alba—White Willow.
„ *viminalis*—Osier.
„ *Caprea*—(Sallow).
Populus tremula—Aspen.

Vanessa urticæ.
Urtica dioica—Stinging Nettle.
„ *urens*—Small Nettle.

Vanessa io.
Urtica dioica—Stinging Nettle.

Vanessa antiopa.
Urtica dioica—Stinging Nettle.
Betula alba—Birch.
Salix alba—White Willow.

Vanessa atalanta.
Urtica dioica—Stinging Nettle.
Parietaria officinalis—Pellitory-of-the-wall.

Vanessa cardui.
Filago germanica—Common Cud-weed.
Carduus nutans—Musk Thistle.
„ *crispus*—Welted Thistle.
Cnicus lanceolatus—Spear Thistle.
„ *arvensis*—Field Thistle.
Onopordon Acanthium—Scotch Thistle.
Echium vulgare—Viper's Bugloss.
Malva sylvestris—Mallow.
Arctium minus (Lappa)—Burdock.

* **Vanessa virginiensis.**
(Food-plant unknown; abroad on *Gnaphalium obtusifolium*.)

* **Danais erippus.**
(Food-plant unknown; in America on *Asclepias*.)

Limenitis sibylla.
Lonicera Periclymenum—Honeysuckle.

Apatura iris.
Salix Caprea—(Sallow).
Populus—Poplar.

Melanargia galatea.
Phleum pratense—Timothy Grass.
Dactylis glomerata—Cock's-foot Grass.
Also on other grasses.

Erebia epiphron (cassiope).
Juncus—Rush.
Aira præcox—Early Hair Grass.
Deschampsia (Aira) flexuosa.
Poa annua—Annual Meadow-grass.
Festuca ovina—Sheep's Fescue-grass.
Nardus stricta—Mat-grass.

Erebia æthiops (medea).
Agrostis canina.
Aira præcox—Early Hair Grass.
Deschampsia (Aira) cæspitosa—Tussock Grass.
Poa annua—Annual Meadow-grass.
„ *pratensis.*

* **Erebia ligea.**
(Food-plant at present unknown.)

Pararge egeria.
Dactylis glomerata—Cock's-foot Grass.
Agropyron (Triticum) repens—Couch Grass.

Pararge megæra.
Dactylis glomerata—Cock's-foot Grass.
Also on other grasses.

Satyrus semele.
Aira præcox—Early Hair Grass.
Deschampsia (Aira) cæspitosa—Tussock Grass.
Agropyron (Triticum) repens—Couch Grass.
Ammophila arundinacea—Marram.

Epinephele ianira (janira).
Poa pratensis.
Also on other grasses.

Epinephele tithonus.
Dactylis glomerata—Cock's-foot Grass.
Poa annua—Annual Meadow-grass.
Agropyron (Triticum) repens—Couch Grass.
Also on various grasses.

Epinephele hyperanthes.
Milium effusum—Millet.
Deschampsia (Aira) cæspitosa—Tussock Grass.
Poa annua—Annual Meadow-grass.
Agropyron (Triticum) repens—Couch Grass.

Cœnonympha (Chortobius) typhon (davus).
Rynchospora alba—White Beak-rush.

Cœnonympha (Chortobius) pamphilus.
Cynosurus cristatus—Dog's-tail.
Poa annua—Annual Meadow-grass.
Nardus stricta—Mat-grass.
Also on other grasses.

Thecla betulæ.
Prunus spinosa—Blackthorn.

Thecla w-album.
Ulmus montana—Wych Elm.
  „   *campestris*—Elm.

Thecla pruni.
Prunus spinosa—Blackthorn.

Thecla quercus.
Quercus Robur—Oak.
Salix (?)—Sallow.

Thecla rubi.
Genista tinctoria—Dyer's Green-weed.
Cytisus (Sarothamnus) scoparius—Broom.
Rubus fruticosus—Bramble.
Betula alba—Birch.

Polyommatus dispar.
Rumex aquaticus.
  „   *Hydrolapathum*—Water Dock.

Polyommatus phlœas.
Senecio Jacobæa—Ragwort.
Rumex pulcher—Fiddle Dock.
  „   *obtusifolius*—Broad-leaved Dock.
  „   *Acetosa*—Sorrel.
  „   *Acetosella*—Sheep Sorrel.

* Polyommatus virgaureæ.
Rumex obtusifolius—Broad-leaved Dock.
  „   *Acetosa*—Sorrel.

Lycæna bætica.
Colutea arborescens—Bladder Senna.   In the pods.
Also on other leguminous plants.

Lycæna ægon.
Ornithopus perpusillus—Bird's-foot.

Lycæna astrarche (medon).
Helianthemum Chamæcistus—Rock Rose.
Erodium cicutarium—Hemlock Stork's-bill.

Lycæna icarus (alexis).
Ononis spinosa—Rest-harrow.
Trifolium pratense—Purple Clover.
„ *repens*—Dutch Clover.
Lotus corniculatus—Bird's-foot Trefoil.
Ornithopus perpusillus—Bird's-foot.

Lycæna bellargus (adonis).
Trifolium repens—Dutch Clover.
Hippocrepis comosa—Horse-shoe Vetch.
Also on other *Leguminosæ*.

Lycæna corydon.
Trifolium pratense—Purple Clover.
„ *repens*—Dutch Clover.
Anthyllis Vulneraria—Lady's Fingers ; Kidney Vetch.
Lotus corniculatus—Bird's-foot Trefoil.
Hippocrepis comosa—Horse-shoe Vetch.

Lycæna argiolus.
Rhamnus catharticus—Buckthorn.
„ *Frangula*—Alder Buckthorn.
Ilex Aquifolium—Holly.
Hedera Helix—Ivy.
Cornus sanguinea—Dog-wood.

Lycæna semiargus (acis).
Melilotus arvensis—Melilot.

Lycæna minima (alsus).
Anthyllis vulneraria—Lady's Fingers ; Kidney Vetch. On the flowers.

Lycæna arion.
Thymus Serpyllum—Thyme.

Lycæna argiades.
Lotus pilosus—Greater Bird's-foot Trefoil.
Also on other *Leguminosæ*.

Nemeobius lucina.
Primula acaulis—Primrose.
„ *veris*—Cowslip.

Syrichthus malvæ (alveolus).
Rubus fruticosus—Bramble.
„ *Idæus*—Raspberry.
Potentilla Fragariastrum—Strawberry-leaved Cinquefoil.

Nisioniades tages.
Lotus corniculatus—Bird's-foot Trefoil.

Hesperia thaumas (linea).
Holcus lanatus—Meadow Soft-grass
Brachypodium sylvaticum.

Hesperia lineola.
On grasses; probably *Holcus lanatus.*

Hesperia actæon.
Calamagrostis epigeios.
Brachypodium sylvaticum.
Agropyron (Triticum) repens—Couch Grass.
„ „ *junceum*—?
„ „ *pungens*—In confinement.

Hesperia sylvanus.
Luzula vernalis (pilosa)—Broad-leaved Wood-rush.
Holcus lanatus—Meadow Soft-grass.
Dactylis glomerata—Cock's-foot Grass.
Agropyron (Triticum) repens.—Couch Grass.

Hesperia comma.
Lotus corniculatus—Bird's-foot Trefoil.
Ornithopus perpusillus—Bird's-foot.
Also on other leguminous plants.

Carterocephalus palæmon (paniscus)
Plantago major—Greater Plantain.

HETEROCERA.

Acherontia atropos.
Solanum Dulcamara—Woody Nightshade.
Lycium barbarum—Tea-tree.
Atropa Belladonna—Deadly Nightshade.
Also on potatoe.

Sphinx convolvuli.
Impatiens Noli-tangere.—Do-not-touch-me.
Calystegia sepium—Great Bindweed.
Convolvulus arvensis—Field Bindweed.
In confinement on lettuce.

Sphinx ligustri.
Ilex Aquifolium—Holly.
Fraxinus excelsior—Ash.
Ligustrum vulgare—Privet.
Syringa vulgaris—Lilac.
　　Also on laurestinus.

Sphinx pinastri.
Pinus—Fir.

Deilephila euphorbiæ.
Euphorbia Cyparissias—Cypress Spurge.
　　　" 　*Paralias*—Sea Spurge.
　　　" 　*Portlandica*—Portland Spurge.
　　　" 　*Peplus*—Petty Spurge.　In confinement.

Deilephila galii.
Rubia peregrina—Madder.
Galium verum—Lady's Bedstraw.
　　" 　*Mollugo*—Hedge Bedstraw.
In confinement on fuchsia.

Deilephila livornica.
Galium verum—Lady's Bedstraw.
Polygonum aviculare—Knot-grass.　In confinement.
Rumex—Dock.　In confinement on this and fuchsia.
Vitis vinifera—Vine.

Chærocampa celerio.
Epilobium hirsutum—Great Hairy Willow herb.
Daucus Carota—Wild Carrot.
Galium Mollugo—Hedge Bedstraw.
Vitis vinifera—Vine.

Chærocampa porcellus.
Galium verum—Lady's Bedstraw.
　　" 　*Mollugo*—Hedge Bedstraw.
　　" 　*palustre*—Water Bedstraw.
Epilobium hirsutum—Great Hairy Willow-herb.
　　" 　*angustifolium*.
Lythrum Salicaria—Purple Loose-strife.
Vitis vinifera—Vine.

Chærocampa elpenor.
Pyrus Malus—Apple.
Epilobium hirsutum—Great Hairy Willow-herb.
Lythrum Salicaria—Purple Loose-strife.
Circæa lutetiana—Enchanter's Night-shade.
Galium verum—Lady's Bedstraw.
　　" 　*Mollugo*—Hedge Bedstraw.　In confinement.
　　" 　*palustre*—Water Bedstraw.
Vitis vinifera—Vine.　In confinement.
In confinement on fuchsia.

* Chærocampa nerii.
  Vitis vinifera—Vine.
  Nerium Oleander—Oleander.
  Vinca major—Greater Periwinkle.

Smerinthus ocellatus.
  Prunus spinosa—Blackthorn.
  Pyrus Malus—Apple.
  Salix alba—White Willow.
  ,,   *cinerea*—(Sallow).
  ,,   *Caprea*—(Sallow).
  Populus alba—White Poplar.
  ,,   *nigra*—Black Poplar.
  ,,   *tremula*—Aspen.

Smerinthus populi.
  Salix cinerea—(Sallow).
  ,,   *Caprea*—(Sallow).
  Populus—Poplar.
  ,,   *nigra (pyramidalis)*—Lombardy Poplar.
  ,,   *tremula*—Aspen.

Smerinthus tiliæ.
  Tilia vulgaris—Lime.
  Ulmus campestris—Elm.
  Corylus Avellana—Hazel.

Macroglossa stellatarum.
  Rubia peregrina—Madder.
  Galium verum—Lady's Bedstraw.
  ,,   *Mollugo*—Hedge Bedstraw.
  ,,   *Aparine*—Goose Grass.

Macroglossa fuciformis.
  Lychnis alba—Evening Campion.
  ,,   *diurna*—Red Campion.
  ,,   *Flos-cuculi*—Ragged Robin.
  Lonicera Periclymenum—Honeysuckle.
  Galium verum—Lady's Bedstraw.
  Scabiosa arvensis—Field Scabious.

Macroglossa bombyliformis.
  Scabiosa succisa—Devil's-bit Scabious.
  ,,   *arvensis*—Field Scabious.

Trochilium apiformis.
  Populus—Poplar. In the stems and wood.
  ,,   *nigra*—Black Poplar. In the bark and wood.
  ,,   *tremula*—Aspen. In the wood.

Trochilium crabroniformis (bembeciformis).
   Salix triandra—Three-stamened Willow. (?) In the wood.
     „ *viminalis*—Osier. In the wood.
     „ *cinerea*—(Sallow). In the wood.
     „ *Caprea*—(Sallow). In the wood.
   Populus nigra (pyramidalis)—Lombardy Poplar. In the bark and wood.
     „ *nigra*—Black Poplar. In the bark and wood.

Sciopteron tabaniformis (asiliformis).
   Fraxinus excelsior—Ash. In the roots.
   Populus nigra—Black Poplar. In the bark and wood.
     „ *tremula*—Aspen. In the wood.

Sesia scoliiformis.
   Betula alba—Birch. In the bark.

Sesia sphegiformis.
   Alnus glutinosa—Alder. In the stems and suckers.

Sesia andreniformis.
   Cornus sanguinea—Dog-wood.
   Ligustrum vulgare—Privet. On the flowers.

Sesia tipuliformis.
   Ribes rubrum—Red Currant. In the shoots.
     „ *nigrum*—Black Currant. In the shoots.

Sesia asiliformis (cynipiformis).
   Ulmus campestris—Elm. In the bark.
   Quercus Robur—Oak. In the bark.

Sesia Myopiformis.
   Pyrus communis—Pear. In the wood.
     „ *Malus*—Apple. In the bark and wood.

Sesia culiciformis.
   Betula alba—Birch. In the bark.
   Alnus glutinosa—Alder. In the bark and wood.

Sesia formiciformis.
   Salix triandra—Three-stamened Willow. In the wood.
     „ *viminalis*—Osier. In the wood.

Sesia Ichneumoniformis.
   Helleborus fœtidus—Stinking Hellebore. In the stems.
   Lotus corniculatus—Bird's-foot Trefoil. In the roots.
   Centaurea nigra—Knapweed. In the stems.

Sesia musciformis (philanthiformis).
   Armeria maritima—Thrift.

Sesia chrysidiformis.
   Rumex—Dock. On the roots.
     „ *Acetosa*—Sorrel. On the roots.
     „ *Acetosella*—Sheep Sorrel. On the roots.

Ino globulariæ.
  Poterium Sanguisorba—Common Salad Burnet.
  Centaurea nigra—Knapweed.
Ino statices.
  Rumex Acetosa—Sorrel.
   ,,   *Acetosella*—Sheep Sorrel.
Ino geryon.
  Helianthemum Chamæcistus—Rock Rose.
  Rumex Acetosa—Sorrel (?).
   ,,   *Acetosella*—Sheep Sorrel.
Zygæna pilosellæ (minos).
  Trifolium ochroleucon.
  Lotus corniculatus—Bird's-foot Trefoil.
  Pimpinella Saxifraga—Common Burnet Saxifrage.
  Thymus Serpyllum—Thyme.
  Hippocrepis comosa—Horse-shoe Vetch.
Zygæna exulans.
  Silene acaulis—Moss Campion.
  Arenaria (Cherleria) sedoides—Mossy Cyphel.
  Medicago lupulina—Black Medick.   In confinement.
  Trifolium pratense—Purple Clover.   In confinement.
     ,,   *repens*—Dutch Clover.   In confinement.
  Lotus corniculatus—Bird's-foot Trefoil.   In confinement.
  Potentilla (Sibbaldia) procumbens—Scotch Cinquefoil.
  Alchemilla alpina—Alpine Lady's Mantle.
  Rumex Acetosa—Sorrel.   In confinement.
Zygæna meliloti.
  Trifolium procumbens—Hop Trefoil.
  Lotus corniculatus—Bird's-foot Trefoil.
Zygæna trifolii.
  Trifolium procumbens—Hop Trefoil.
  Lotus corniculatus—Bird's-foot Trefoil.
  Hippocrepis comosa—Horse-shoe Vetch.
Zygæna loniceræ.
  Trifolium pratense—Purple Clover.
     ,,   *repens*—Dutch Clover.
  Lotus corniculatus—Bird's-foot Trefoil.
  Lathyrus (Orobus) pratensis—Meadow Vetchling.
  Hippocrepis comosa—Horse-shoe Vetch.
Zygæna filipendulæ.
  Trifolium pratense—Purple Clover.
     ,,   *repens*—Dutch Clover.
  Lotus corniculatus—Bird's-foot Trefoil.
  Onobrychis viciæfolia (sativa)—Saint-foin.
  Valeriana officinalis—Great Wild Valerian.

* Syntomis phegea.
   Scabiosa succisa—Devil's-bit Scabious.
   Taraxacum officinale—Dandelion.
   Plantago lanceolata—Narrow-leaved Plantain.
   Rumex Acetosa—Sorrel.

* Nacilia ancilla.
   Lichen. On trees.

Sarothripus undulanus (revayana).
   Salix cinerea—(Sallow).
     „ *Caprea*—(Sallow).

Earias chlorana.
   Salix triandra—Three-stamened Willow (?). In the shoots.
     „ *viminalis*—Osier (?). In the shoots.
     „ *alba*—White Willow. In the shoots.

Hylophila (Halias) prasinana.
   Quercus Robur—Oak.
   Corylus Avellana—Hazel.
   Fagus sylvatica—Beech.
   Betula alba—Birch.
   Alnus glutinosa—Alder.

Hylophila (Halias) bicolorana (quercana).
   Quercus Robur—Oak.

Nola cucullatella.
   Prunus spinosa—Blackthorn.
   Prunus—Plum.
   Cratægus Oxyacantha—Whitethorn.

Nola strigula.
   Quercus Robur—Oak.
   Lichen caninus. On oaks.

Nola confusalis (cristulalis).
   Quercus Robur—Oak.
   Fagus sylvatica—Beech.

Nola albulalis.
   Rubus cæsius—Dewberry.

Nola centonalis.
   Medicago lupulina—Black Medick.
   Trifolium pratense—Purple Clover. On the flowers.
     „ *dubium*—On the flowers.
     „ *repens*—Dutch Clover. On the flowers.
     „ *procumbens*—Hop Trefoil. Its proper food.
     „ *filiforme.*
   Lotus corniculatus—Bird's-foot Trefoil.
   Potentilla Anserina—Silver-weed.

Nudaria senex.
Lichen.
Lichen caninus. In confinement on this and decayed sallow or bramble leaves.

Nudaria mundana.
Lichen. On trees and walls.

Setina irrorella.
Lichen. On stones on the sea-coast above high-water.

Calligenia miniata.
Lichen. On oaks, beech and birch.
Lichen caninus. On oaks.
In confinement on withered oak or sallow leaves.

Lithosia mesomella.
Lichen. On oaks.
In confinement on green or withering sallow leaves.

Lithosia muscerda.
Lichen. Probably on sallows in wet pla es.
In confinement on *Lichen caninus* and decayed sallow or bramble leaves.

Lithosia sororcula (aureola).
Lichen. On oaks, beeches, larch, fir and pine.

Lithosia lutarella (pygmæola).
Lichen.

Lithosia griseola.
Lichen.
  „   *caninus.*
Trifolium repens—Dutch Clover.
  „   *pratense*—Purple Clover.
Polygonum aviculare—Knot-grass. In confinement.
In confinement on mosses and withered sallow leaves.

Lithosia griseola v. stramineola.
Lichen.
Lichen caninus.

Lithosia deplana.
Lichen. On oaks, beech and yew.

Lithosia lurideola (complanula).
Clematis vitalba—Traveller's Joy.
Lichen (?). On oaks, blackthorn, poplars, larch and walls.
Rhamnus catharticus—Buckthorn.
Rhamnus Frangula—Alder Buckthorn.
Cornus sanguinea—Dog-wood.
Quercus Robur—Oak.

Lithosia complana.
Lichen. On blackthorn and fir.

Lithosia sericea (molybdeola).
Lichen. Probably on heath or stones underneath. In confinement on chickweed, dandelion, lettuce, and withered oak or sallow leaves.

Lithosia caniola.
Trifolium pratense—Purple Clover.
Trifolium repens—Dutch Clover.
Lotus corniculatus—Bird's-foot Trefoil.
Also on other *Leguminosæ*.

Gnophria quadra.
Lichen—On trees.
Lichen caninus—On oaks.

Gnophria rubricollis.
Lichen—On beech and other trees.

Emydia cribrum.
Calluna Erica—Ling; Heather.

Emydia striata. (grammica.)
Artemisia vulgaris—Mugwort. On the seeds.
Calluna Erica—Ling; Heather.
Festuca ovina—Sheep's Fescue-grass.

Deiopeia pulchella.
Myosotis palustris—Forget-me-not. In confinement.
„ *arvensis*—Field Forget-me-not.
Borago officinalis—Borage. In confinement.

Euchelia jacobææ.
Senecio vulgaris—Groundsel.
„ *Jacobæa*—Ragwort.

Callimorpha dominula.
Spiræa Ulmaria—Meadow-sweet.
Cynoglossum officinale—Hound's-tongue.
Salix repens—Dwarf Sallow.
Also on various low plants.

Callimorpha hera.
Cynoglossum officinale—Hound's-tongue.
Lamium album—White Dead-nettle.
Plantago—Plantain.
Taraxacum officinale—Dandelion.
Also on other low plants.

Nemeophila russula.
   Hieracium Pilosella—Mouse-ear Hawk-weed.
   Taraxacum officinale—Dandelion.
   Erica cinerea—Common Purple Heath.
   Plantago—Plantain. In confinement on lettuce.
   Also on other low plants.

Nemeophila plantaginis.
   Viola odorata—Sweet Violet.
     „ *canina*—Dog Violet.
     „ *tricolor*—Heartsease.
   Senecio vulgaris—Groundsel.
   Plantago—Plantain.
   Also on numerous low plants.

Arctia caja.
   Tilia vulgaris—Lime.
   Pyrus Malus—Apple.
   Lamium purpureum—Purple Dead-nettle.
     „ *album*—White Dead-nettle.
   Symphytum officinale—Comfrey.

Arctia villica.
   Stellaria media—Chickweed. In confinement on lettuce.
   Also on numerous low plants.

Spilosoma fuliginosa.
   Plantago—Plantain.
   Listera ovata—Twayblade.

Spilosoma mendica.
   Stellaria media—Chickweed.
   Plantago—Plantain.
   Rumex—Dock.
   Urtica dioica—Stinging Nettle.
   Betula alba—Birch.
   Also on various low plants.

Spilosoma lubricepeda.
   Cratægus Oxyacantha—Whitethorn.
   Rumex—Dock.
   Also on very many low plants.

Spilosoma menthastri.
   Rumex—Dock.
   Urtica dioica—Stinging Nettle.
   Also on very many low plants.

Spilosoma urticæ.
   Epilobium hirsutum—Great Hairy Willow-herb.
   Pedicularis palustris—Marsh Red Rattle.
   Mentha hirsuta—Hairy Mint. (?)
   Iris Pseudacorus—Yellow Iris.

LARVÆ AND FOOD-PLANTS.

Hepialus humuli.
Arctium minus (Lappa)—Burdock. On the roots.
Taraxacum officinale—Dandelion. On the roots of this and grasses.
Scrophularia aquatica—Water Fig-wort.
Lamium album—White Dead-nettle. On the roots.
Ballota nigra—Black Horehound. On the roots.
Rumex—Dock. On the roots.
Humulus Lupulus—Hop. On the roots.

Hepialus sylvanus.
Rumex—Dock. On the roots.

Hepialus velleda.
Pteris aquilina—Brake; Bracken. On the roots.

Hepialus lupulinus.
Lamium purpureum—Purple Dead-nettle. On the roots.
 " *album*—White Dead-nettle. On the roots.
Ballota nigra—Black Horehound. On the roots.

Hepialus hectus.
Taraxacum officinale—Dandelion. On the roots of this and other plants.
Pteris aquilina—Brake; Bracken. On the roots.

Cossus ligniperda.
Fraxinus excelsior—Ash. In the wood.
Quercus Robur—Oak. In the wood.
Salix alba—White Willow. In the wood.
Populus—Poplar. In the wood.
Syringa vulgaris—Lilac. In the wood.

Zeuzera pyrina (æsculi).
Acer Pseudo-platanus—Sycamore. In the wood.
Prunus—Plum. In the wood.
Pyrus communis—Pear. In the wood.
 " *Malus*—Apple. In the wood.
Cratægus Oxyacantha—Whitethorn. In the wood.
Fraxinus excelsior—Ash. In the wood.
Ulmus campestris—Elm. In the wood.
Betula alba—Birch. In the wood.
Alnus glutinosa—Alder. In the wood.
Populus—Poplar. In the wood.
Æsculus hippocastanum—Horse Chestnut. In the wood.
Syringa vulgaris—Lilac. In the wood.

Macrogaster castaneæ (arundinis).
Phragmites communis—Reed. In the stems.

Heterogenea limacodes (testudo).
Betula alba—Birch.
Quercus Robur—Oak.
Fagus sylvatica—Beech.

Heterogenea asella (asellus.)
Betula alba—Birch.
Quercus Robur—Oak.
Fagus sylvatica—Beech.
Populus—Poplar.

Porthesia (Liparis) chrysorrhœa.
Prunus spinosa—Blackthorn.
Cratægus Oxyacantha—Whitethorn.
Also on standard roses.

Porthesia (Liparis) similis (auriflua).
Pyrus Malus—Apple.
Cratægus Oxyacantha—Whitethorn.
Quercus Robur—Oak.

Leucoma salicis.
Salix alba—White Willow.
Populus—Poplar.
  „   *nigra (pyramidalis)*.—Lombardy Poplar.

Lælia cœnosa.
Cladium germanicum—Twig-rush.
Phragmites communis—Reed.
Carex—Sedge.

Ocneria dispar.
Prunus spinosa—Blackthorn.
  „   —Plum.
Pyrus Malus—Apple.
Cratægus Oxyacantha—Whitethorn.

*Laria l-nigrum.
Tilia vulgaris—Lime.
Quercus Robur—Oak.
Fagus sylvatica—Beech.
Salix Caprea—(Sallow).

Psilura monacha.
Pyrus Malus—Apple.
Betula alba—Birch.
Quercus Robur—Oak.
Pinus—Fir.

Dasychira fascelina.
Prunus—Plum.
Corylus Avellana—Hazel.
Also on various trees and herbaceous plants.

Dasychira pudibunda.
Tilia vulgaris—Lime.
Betula alba—Birch.
Quercus Robur—Oak.
Castanea sativa—Chestnut.
Humulus Lupulus—Hop.
Also on various trees.

Orgyia gonostigma.
Prunus spinosa—Blackthorn.
Cratægus Oxyacantha—Whitethorn.
Corylus Avellana—Hazel.
Quercus Robur—Oak.
Betula alba—Birch.
Salix—(Sallow).

Orgyia antiqua.
Tilia vulgaris—Lime.
Prunus spinosa—Blackthorn.
Prunus—Plum.
Pyrus communis—Pear.
 „ *Malus*—Apple.
Cratægus Oxyacantha—Whitethorn.
Corylus Avellana—Hazel.
Quercus Robur—Oak.
Salix cinerea—(Sallow). In confinement.
 „ *Caprea*—(Sallow). In confinement.
Robinia Pseud-acacia—Acacia.
Also on many trees and shrubs.

Trichiura cratægi.
Prunus spinosa—Blackthorn.
Cratægus Oxyacantha—Whitethorn.
Betula alba—Birch.
Quercus Robur—Oak.
Salix cinerea—(Sallow).
 „ *Caprea*—(Sallow).
Populus nigra—Black Poplar.

Pœcilocampa populi.
Tilia vulgaris—Lime,
Cratægus Oxyacantha—Whitethorn.
Fraxinus excelsior—Ash.
Betula alba—Birch.
Quercus Robur—Oak.
Populus—Poplar.

Eriogaster lanestris.
Prunus spinosa—Blackthorn.
Cratægus Oxyacantha—Whitethorn.
Ulmus campestris—Elm.

Bombyx neustria.
Prunus—Plum.
Pyrus communis—Pear.
  „ *Malus*—Apple.
Corylus Avellana—Hazel.

Bombyx castrensis.
Daucus Carota—Wild Carrot.
Artemisia maritima—Sea Wormwood.
Plantago lanceolata—Narrow-leaved Plantain.
In confinement on apple, pear, rose, birch, sprinkled with salt-water.

Bombyx rubi.
Rubus fruticosus—Bramble.
Rosa spinosissima—Burnet Rose.
Vaccinium Myrtillus—Whortleberry; Bilberry.
Calluna Erica—Ling; Heather.
Erica Tetralix—Cross-leaved Heath.
  „ *cinerea*—Common Purple Heath.
Polygonum aviculare—Knot-grass.

Bombyx quercus.
Cratægus Oxyacantha—Whitethorn.
Calluna Erica—Ling; Heather.
Populus nigra—Black Poplar.

Bombyx quercus v. callunæ.
Cratægus Oxyacantha—Whitethorn.

Bombyx trifolii.
Cytisus (Sarothamnus) scoparius—Broom.
Medicago sativa—Lucerne.
Melilotus arvensis—Melilot.
Trifolium pratense—Purple Clover.
  „ *repens*—Dutch Clover.

Odonestis potatoria.
Poa annua—Annual Meadow-grass.
Also on other grasses.

Lasiocampa quercifolia.
Rhamnus catharticus—Buckthorn.
Prunus spinosa—Blackthorn.
Salix alba—White Willow.

Lasiocampa ilicifolia.
Vaccinium Myrtillus—Whortleberry; Bilberry.
Salix cinerea—(Sallow).
  „ *Caprea*—(Sallow).

Endromis versicolor.
Betula alba—Birch.

Saturnia pavonia (carpini).
Prunus spinosa—Blackthorn.
Spiræa Ulmaria—Meadow-sweet.
Sambucus nigra—Elder.
Erica Tetralix—Cross-leaved Heath.
" *cinerea*—Common Purple Heath.
Salix alba—White Willow.
Cratægus Oxyacantha—Whitethorn.
Rubus fruticosus—Bramble.
Salix—Sallow.
Lythrum Salicaria—Purple Loosestrife.
In confinement on walnut.

Drepana (Platypteryx) lacertinaria (lacertula).
Betula alba—Birch.

Drepana (Platypteryx) harpagula (sicula).
Quercus Robur—Oak.
Betula alba—Birch.
Tilia vulgaris—Lime.
" *cordata (parvifolia)*.

Drepana (Platypteryx) falcataria (falcula).
Populus tremula—Aspen.
Salix alba—White Willow.
Quercus Robur—Oak.
Alnus glutinosa—Alder.
Betula alba—Birch.

Drepana (Platypteryx) binaria (hamula).
Quercus Robur—Oak.
Betula alba—Birch.

Drepana (Platypteryx) cultraria (unguicula).
Fagus sylvatica—Beech.

Cilix glaucata (spinula).
Cratægus Oxyacantha—Whitethorn.
Rhamnus spinosa—Blackthorn.
Pyrus communis—Pear.

Dicranura bicuspis.
Betula alba—Birch.
Alnus glutinosa—Alder.
Fagus sylvatica—Beech.

Dicranura furcula.
Fagus sylvatica.—Beech.
Salix alba—White Willow.
" *viminalis*—Osier.
" *cinerea*—(Sallow.
" *Caprea*—(Sallow.)

Dicranura bifida.
   Populus—Poplar.
    „   *tremula*—Aspen.

Dicranura vinula.
   Salix alba—White Willow.
    „ *cinerea*—(Sallow.)
    „ *Caprea*—(Sallow.)
   Populus—Poplar.
   Populus nigra—Black Poplar.

Stauropus fagi.
   Pyrus Malus—Apple.
   Betula alba—Birch.
   Quercus Robur—Oak.
   Fagus sylvatica—Beec

Glyphisia crenata.
   Populus—Poplar.
   Populus nigra—Black Poplar.

Ptilophora plumigera.
   Acer campestre—Maple.
    „ *Pseudo-platanus*—Sycamore.

Pterostoma (Ptilodontis) palpina.
   Salix cinerea—(Sallow.)
    „ *Caprea*—(Sallow.)
   Populus—Poplar.
   Populus tremula—Aspen.

Lophopteryx camelina.
   Acer campestre—Maple.
   Betula alba—Birch.
   Alnus glutinosa—Alder.
   Quercus Robur—Oak.
   Populus—Poplar.
   Corylus Avellana—Hazel.

Lophopteryx cuculla (cucullina).
   Acer Pseudo-platanus—Sycamore.
    „ *campestre*—Maple.

Lophopteryx carmelita.
   Betula alba—Birch.

Notodonta bicolor.
   Betula alba—Birch.

Notodonta dictæa.
   Salix alba—White Willow.
    „ *Caprea*—(Sallow.)
   Populus nigra—Black Poplar.
    „ *tremula*—Aspen.

Notodonta dictæoides.
Acer Pseudo-platanus—Sycamore.
Betula alba—Birch.
Salix Caprea—(Sallow.)

Notodonta dromedarius.
Betula alba—Birch.
Alnus glutinosa—Alder.

Notodonta trilophus.
Betula alba—Birch.
Populus—Poplar.
  ,,   *tremula*—Aspen.

Notodonta ziczac.
Salix alba—White Willow.
  ,, *cinerea*—(Sallow.)
  ,, *Caprea*—(Sallow.)
Populus—Poplar.

Notodonta trepida.
Quercus Robur—Oak.

Notodonta chaonia.
Quercus Robur—Oak.

Notodonta trimacula (dodonea).
Betula alba—Birch.
Quercus Robur—Oak.

Phalera bucephala.
Tilia vulgaris—Lime.
Prunus—Plum.
Ulmus campestris—Elm.
Corylus Avellana—Hazel.
Quercus Robur—Oak.
Also on other trees.

Pygæra (Clostera) curtula.
Salix cinerea—(Sallow.)
  ,, *Caprea*—)Sallow.)
Populus—Poplar.
  ,, *tremula*—Aspen.

Pygæra (Clostera) anachoreta.
Salix Caprea—(Sallow.)
Populus nigra—Black Poplar.

Pygæra (Clostera) pigra (reclusa).
Salix cinerea—(Sallow.)
  ,, *Caprea*—(Sallow.)
Populus tremula—Aspen.

Thyatira derasa.
Rubus fruticosus—Bramble.

Thyatira batis.
Rubus fruticosus—Bramble.

Cymatophora octogesima (ocularis).
Populus tremula—Aspen.

Cymatophora or.
Populus tremula—Aspen.
　　,,　*nigra*—Black Poplar.

Cymatophora duplaris.
Betula alba—Birch.
Alnus glutinosa—Alder.

Cymatophora fluctuosa.
Betula alba—Birch.

Asphalia diluta.
Betula alba—Birch.
Quercus Robur—Oak.

Asphalia flavicornis.
Betula alba—Birch.

Asphalia ridens.
Quercus Robur—Oak.

Bryophila algæ.
Lichen. On trees.

Bryophila muralis (glandifera.)
Lichen. On walls and buildings.

Bryophila par (var. of above.)
Lichen. On walls and buildings.

Bryophila perla.
Lichen. On walls.

Moma (Diphthera) orion.
Betula alba—Birch.
Quercus Robur—Oak.

Demas coryli.
Prunus spinosa—Blackthorn.
Corylus Avellana—Hazel.
Quercus Robur—Oak.

Acronycta tridens.
Prunus spinosa—Blackthorn.
Pyrus communis—Pear.
Cratægus Oxyacantha—Whitethorn.
Also on various trees, shrubs and plants.

Acronycta psi.
Tilia vulgaris—Lime.
Prunus spinosa—Blackthorn.
  „   —Plum.
Pyrus communis—Pear.
  ,.   *Malus*—Apple.
Cratægus Oxyacantha—Whitethorn.
Also on various trees, shrubs and plants.

Acronycta leporina.
Betula alba—Birch.
Alnus glutinosa—Alder.
Salix alba—White Willow.
  „   *cinerea*—(Sallow.)
  „   *Caprea*—(Sallow.)

Acronycta aceris.
Acer Pseudo-platanus—Sycamore.
Quercus Robur—Oak.

Acronycta megacephala.
Populus—Poplar.

Acronycta strigosa.
Prunus spinosa—Blackthorn.
Cratægus Oxyacantha—Whitethorn.

Acronycta alni.
Tilia vulgaris—Lime.
Rubus Idæus—Raspberry.
Rosa canina—Dog Rose.
Cratægus Oxyacantha—Whitethorn.
Betula alba—Birch.
Alnus glutinosa—Alder.
Pyrus Malus—Apple.
Ulmus montana—Mountain Ash.
Corylus Avellana—Hazel.
Quercus Robur - Oak.
Castanea sativa—Chestnut.
Fagus sylvatica—Beech.
Salix alba—White Willow.
  „   *cinerea* (Sallow.)
  „   *Caprea*—(Sallow.)

Acronycta ligustri.
Fraxinus excelsior—Ash.
Ligustrum vulgare—Privet.

Acronycta rumicis.
Fragaria vesca—Wild Strawberry.
Calluna Erica—Ling; Heather.
Also on many low plants.

Acronycta auricoma.
Rubus fruticosus—Bramble.
Vaccinium Myrtillus—Whortleberry; Bilberry.
Polygonum aviculare—Knot-grass.
Betula alba—Birch.
Quercus Robur—Oak.
Also on many other plants.

Acronycta menyanthidis.
Calluna Erica—Ling; Heather.
Cratægus Oxyacantha—Whitethorn. In confinement.
Myrica Gale—Sweet Gale.
Salix alba—White Willow.
Menyanthes trifoliata—Buck-bean; Marsh Trefoil.

Acronycta euphorbiæ.
Myrica Gale—Sweet Gale.
Salix—(Sallow.) In confinement.
Calluna Erica—Ling; Heather. In confinement.
Betula alba—Birch. In confinement.
Erica—Heath. In confinement.

Diloba cæruleocephala.
Prunus—Plum.
Pyrus communis—Pear.
Cratægus Oxyacantha—Whitethorn.

Arsilonche albovenosa (venosa).
Carex—Sedge.
Phragmites communis—Reed.
Glyceria aquatica.
Also on various water plants.

Synia musculosa.
(Food-plant at present unknown.)

Leucania conigera.
Agropyron (Triticum) repens—Couch-grass.
Also on other grasses.

Leucania vitellina.
Nardus stricta—Mat-grass (?).
Also on other grasses.

Leucania turca.
Luzula vernalis (pilosa)—Broad-leaved Wood-rush.
   ,,   *campestris*—Field Wood-rush; Cuckoo-grass.

Leucania lithargyria.
Stellaria media—Chickweed.
Plantago—Plantain.
Also on grasses.

Leucania albipuncta.
Stellaria media—Chickweed.
Agropyron (Triticum) repens—Couch-grass.

Leucania extranea.
(Food-plant at present unknown.)

Leucania obsoleta.
Phragmites communis—Reed.

Leucania putrescens.
Dactylis glomerata—Cock's-foot Grass (?).
On coast grasses.

Leucania littoralis.
Carex riparia.
Carex sylvatica—In confinement.
Ammophila arundinacea—Marram.

Leucania impudens (pudorina).
Phragmites communis—Reed.
Also on several grasses.

Leucania comma.
Rumex Acetosa—Sorrel.
Rumex Acetosella—Sheep Sorrel.
Dactylis glomerata—Cock's-foot Grass.
Also on other grasses.

Leucania straminea.
Poa trivialis—(?)
Phragmites communis—Reed.
Phalaris arundinacea—Reed Canary Grass.
Also on other grasses "in damp meadows and on the banks of streams."

Leucania impura.
Carex—Sedge.

Leucania pallens.
Deschampsia (Aira) cæspitosa—Tussock-grass.
Poa annua—Annual Meadow-grass.
Agropyron (Triticum) repens—Couch-grass.
Dactylis glomerata—Cock's-foot Grass.

*** Leucania loreyi.**
(Grass—species unknown.)

*** Leucania l-album.**
(Grass—species unknown.)

Calamia phragmitidis.
Phragmites communis—Reed.

Meliana flammea.
Phragmites communis—Reed. In the stems.

Senta maritima (ulvæ).
Phragmites communis—Reed. In the stems.

Cœnobia rufa (despecta).
Juncus lamprocarpus. In the stems.

Tapinostola fulva.
Carex—Sedge. In the stems.
„ *paludosa.* In the stems.
Eriophorum angustifolium—Common Cotton Grass.
Glyceria aquatica.

Tapinostola hellmanni.
Phragmites communis—Reed.

Tapinostola extrema (concolor).
(Food-plant at present unknown.)

Tapinostola bondii.
Arrhenatherum avenaceum.

Tapinostola elymi.
Elymus arenarius—Lyme-grass. In the stems and roots.

Nonagria cannæ.
Typha latifolia—Reed Mace.

Nonagria sparganii.
Iris Pseudacorus—Yellow Iris. In the stems of this and other Iris.
Typha latifolia—Reed Mace. In the stems.
„ *angustifolia.* In the stems.
Sparganium ramosum—Bur-reed. In the stems.

Nonagria arundinis (typhæ).
Typha latifolia—Reed Mace. In the stems.

Nonagria geminipuncta.
Phragmites communis—Reed. In the stems.

Nonagria neurica.
Phragmites communis—Reed. In the stems.

Nonagria brevilinea.
Typha latifolia—Reed Mace.
Phragmites communis—Reed. In the stems.
Scirpus. In the stems.

Nonagria lutosa.
Phragmites communis—Reed. In the stems.

Gortyna ochracea (flavago).
   Arctium minus (Lappa)—Burdock. In the stems.
   Carduus nutans—Musk Thistle. In the stems of this and other thistles.
   Cnicus palustris—Marsh Thistle. In the stems.
   Verbascum Thapsus—Mullein. In the stems.
   Scrophularia aquatica—Water Fig-wort. In the stems.
   Digitalis purpurea—Foxglove. In the stems.
   Sambucus nigra—Elder.
   Also on potatoe.

Hydræcia nictitans.
   Poa annua—Annual Meadow-grass. On the roots of this and other grasses.
   Glyceria maritima.

Hydrœcia petasitis.
   Petasites officinalis (vulgaris)—Butter-Bur. In the stems and roots.

Hydrœcia micacea.
   Cyperus longus—Cypress-root.
   Carex—Sedge.
   Equisetum arvense—Field-Mare's Tail. In the stems.
   „   *limosum.* In the stems.

Axylia putris.
   Primula acaulis—Primrose (?).
   Humulus Lupulus—Hop.
   Urtica dioica—Stinging Nettle.
   Also on various low plants.

Xylophasia rurea.
   Primula acaulis—Primrose.
   „   *veris*—Cowslip.
   Also on various grasses.

Xylophasia lithoxylea.
   Poa annua—Annual Meadow-grass (?). On the roots of this and other grasses.

Xylophasia sublustris.
   Poa annua—Annual Meadow-grass (?). On the roots of this and other grasses.

Xylophasia monoglypha (polyodon).
   Poa annua—Annual Meadow-grass (?). On the roots of this and other grasses and low plants.

Xylophasia hepatica.
   Stellaria media—Chickweed.
   Also on grasses and roots of other low plants.

Xylophasia scolopacina.
Luzula vernalis (pilosa) Broad-leaved Wood-rush.
" *campestris*—Field Wood-rush ; Cuckoo-grass.
Scirpus cæspitosus.
Briza media—Quaking Grass ; Maidenhair.
" *minor*—Small Quaking Grass.

Dipterygia scabriuscula (pinastri).
Rumex—Dock.

Cloantha polyodon (perspicillaris).
Hypericum perforatum—Perforated St. John's Wort

Aporophyla australis.
Cichorium Intybus—Chicory.
Poa annua—Annual Meadow-grass.
Stellaria media—Chickweed.

Laphygma exigua.
Plantago—Plantain.

Neuria reticulata (sapponariæ).
Silene Cucubalus—White Campion.
" *maritima*—Sea Campion.
" *gallica*—English Catchfly.
Also on other low plants.

Neuronia popularis.
Poa annua—Annual Meadow-grass.
Also on other grasses.

Heliophobus hispidus.
Plantago—Plantain.
Poa annua—Annual Meadow-grass.
Also on other grasses.

Charæas graminis.
Poa annua—Annual Meadow-grass. On the roots of this and other grasses.

Pachetra leucophæa.
Poa annua—Annual Meadow-grass.
" *nemoralis*—Wood Meadow-grass.
Festuca ovina—Sheep's Fescue Grass (?.)
" In tufts of grass growing in woods," and on commons.

Cerigo matura (cytherea).
Aira caryophyllea—Hair-grass.
Poa annua—Annual Meadow-grass.
Also on other grasses on high and stony places.

Luperina testacea.
Poa annua—Annual Meadow-grass. On the stems of this and other grasses.

Luperina dumerili.
(Food-plant at present unknown).

Luperina cespitis.
Poa annua—Annual Meadow-grass.
Also on other grasses.

Mamestra abjecta.
Artemisia maritima—Sea Wormwood.

Mamestra sordida (anceps).
(Grass; species at present unknown).

Mamestra albicolon.
Chenopodium Vulvaria—Stinking Goose-foot; Orach.
,,   *album*—White Goose-foot.
,,   *Bonus-Henricus*—Good King Henry.
Atriplex patula—Orach.
,,   *laciniata*—Sea Orach.
In gardens on lettuce.

Mamestra furva.
Corynephorus (Aira) canescens.

Mamestra brassicæ.
Chenopodium Vulvaria—Stinking Goose-foot; Orach.
,,   *album*—White Goose-foot.
Bonus-Henricus—Good King Henry.
Rumex—Dock.
Also on many other plants.

Mamestra persicariæ.
Sambucus nigra—Elder.
Also on low plants.

Apamea basilinea.
Agropyron (Triticum)—Wheat. On the grain, afterwards on various low plants.

Apamea connexa.
Grass in woods; species unknown. In the stems (?).

Apamea gemina.
Poa annua—Annual Meadow-grass. In confinement.
Agropyron (Triticum) repens—Couch-grass. In confinement.
Phalaris arundinacea—Reed Canary Grass. In confinement.

Apamea unanimis.
Agropyron (Triticum) repens—Couch-grass. In confinement.
Phalaris arundinacea—Reed Canary Grass.
,,   *canariensis*—Canary Grass.
Carex—Sedge.

Apamea ophiogramma.
  Phalaris arundinacea—Reed Canary Grass.
  „ *canariensis*—Canary Grass.
  Iris Pseudacorus—Yellow Iris. In the stems.
Apamea leucostigma (fibrosa).
  Iris Pseudacorus—Yellow Iris. In the stems.
  Cladium germanicum—Twig-rush.
Apamea didyma (oculea).
  Poa annua—Annual Meadow-grass.
  Also on other grasses.
Miana strigilis.
  Dactylis glomerata—Cock's-foot Grass.
Miana fasciuncula.
  Deschampsia (Aira) cæspitosa—Tussock-grass. In the stems of this and other grasses.
Miana literosa.
  Poa annua—Annual Meadow-grass (?). In the stems of this and other grasses.
Miana bicoloria (furuncula.)
  Arrhenatherum avenaceum.
Miana arcuosa.
  Deschampsia (Aira) cæspitosa—Tussock-grass. On the roots.
Phothedes captiuncula.
  Carex glauca.
  Phalaris arundinacea—Reed Canary Grass.
Celæna haworthii.
  Eriophorum angustifolium—Common Cotton Grass.
  „ *vaginatum*—Hare's-tail Cotton Grass.
Grammesia trigrammica (trilinea).
  Plantago major—Greater Plantain.
Stilbia anomala.
  Poa annua—Annual Meadow-grass.
  Also on other grasses.
Caradrina morpheus.
  Dipsacus sylvestris—Wild Teasel.
  „ *pilosus*—Small Teasel.
  Chenopodium Vulvaria—Stinking Goose-foot; Orach.
  „ *album*—White Goose-foot.
  Sedum Telephium—Orpine.
  Galium Mollugo—Hedge Bedstraw.
  Salix—(Sallow.)
  Humulus Lupulus—Hop.
  Also on other plants.

Caradrina alsines.
>*Stellaria media*—Chickweed.
>*Plantago*—Plantain.
>Also on various low plants.

Caradrina taraxaci (blanda.)
>*Stellaria media*—Chickweed.
>Also on various low plants.

Caradrina ambigua.
>(Food-plant at present unknown.)

Caradrina quadripunctata (cubicularis).
>*Stellaria media*—Chickweed.
>Also on various farinaceous and leguminous plants.

Acosmetia caliginosa.
>*Poterium Sanguisorba*—Common Salad Burnet.

Hydrilla palustris.
>*Plantago*—Plantain.
>Also on various low plants.

Rusina tenebrosa.
>*Viola odorata*—Sweet Violet.
>„ *canina*—Dog Violet.
>„ *tricolor*—Heartsease.
>Also on various low plants.

Agrotis vestigialis (valligera.)
>*Holcus lanatus*—Meadow Soft-grass. (?) On the roots of this and other grasses.
>*Borago officinalis*—Borage.
>*Artemisia campestris*—Field Wormwood.

Agrotis puta.
>*Taraxacum officinale*—Dandelion. (?)
>*Polygonum aviculare*—Knot-grass.
>In confinement on carrot and lettuce.

Agrotis suffusa.
>*Cichorium Intybus*—Chicory.
>*Holcus lanatus*—Meadow Soft-grass. (?) On the roots of this and other grasses.

Agrotis saucia.
>*Trifolium pratense*—Purple Clover.
>„ *repens*—Dutch Clover.
>*Plantago*—Plantain.
>*Rumex*—Dock.
>Also on other low plants.

Agrotis segetum.
    Brassica Sinapis—Cherlock; Wild Mustard. On the roots of this, turnip, cabbage, horse-radish, carrot, mangold-wurzel, grasses, china asters, etc.

Agrotis lunigera.
    Polygonum aviculare—Knot-grass. In confinement.
    Taraxacum officinale—Dandelion. In confinement.

Agrotis exclamationis.
    Brassica Napus—Rape. On the roots of this, turnip, other vegetables and low plants.

Agrotis corticea.
    Chenopodium album—White Goose-foot. In confinement.
    Polygonum aviculare—Knot-grass. In confinement.
    Rumex—Dock. In confinement.
    Verbascum—Mullein. In confinement.
    Also on hollyhock and slices of carrot.

Agrotis cinerea.
    Rumex Acetosa—Sorrel. On the roots of this and other low plants.
    Rumex—Dock. In confinement; prefers the seeds.

Agrotis ripæ.
    Cynoglossum officinale—Hound's-tongue.

Agrotis cursoria.
    Euphorbia Esula.
    Arenaria peploides—Sea Sandwort.
    Viola Curtisii Forster.
    Agropyron (Triticum) junceum.

Agrotis nigricans.
    Trifolium pratense—Purple Clover.
        „   *repens*—Dutch Clover.
    Also on various low plants.

Agrotis tritici.
    Plantago maritima—Sea Plantain. On the roots of this and other low plants and vegetables.

Agrotis aquilina.
    Papaver Rhœas—Red Poppy. On the roots.
    Stellaria media—Chickweed. On the roots.
    Galium verum—Lady's Bedstraw.
    Plantago—Plantain.
    Also on the roots of other poppies, cabbage, brocoli, etc.

Agrotis obelisca.
    Galium verum—Lady's Bedstraw.
    Also on various low plants.

Agrotis agathina.
Calluna Erica—Ling ; Heather.

Agrotis strigula (porphyrea).
Calluna Erica—Ling ; Heather.

Agrotis præcox.
Stellaria media—Chickweed.
Also on various low plants.

Agrotis obscura (ravida).
Carduus nutans—On the roots of this and other thistles.
Taraxacum officinale—Dandelion. On the roots and leaves.
Rumex—Dock.

Agrotis simulans (pyrophila).
Polygonum aviculare—Knot-grass. (?)
Also on grasses and other low plants.

Agrotis lucernea.
Taraxacum officinale—Dandelion.
Campanula rotundifolia—Harebell. In confinement.
Also on other low plants.

Agrotis ashworthii.
Helianthemum Chamæcistus—Rock Rose.
Poterium Sanguisorba—Common Salad Burnet.
Scabiosa succisa—Devil's-bit Scabious.
Solidago Virgaurea—Golden-rod.
Hieracium Pilosella—Mouse-ear Hawk-weed.
Campanula rotundifolia—Harebell. In confinement.
Calluna Erica—Ling ; Heather. In confinement.
Thymus Serpyllum—Thyme.
Polygonum aviculare—Knot-grass.
Salix Caprea—(Sallow). In confinement.
Festuca ovina—Sheep's Fescue-grass.
Also on other *Hieracia*.

Noctua glareosa.
Cytisus (Sarothamnus) scoparius—Broom.
Rumex—Dock.
   ,, *Acetosa*—Sorrel.
   ,, *Acetosella*—Sheep Sorrel.

Noctua depuncta.
Rumex Acetosa—Sorrel.
   ,, *Acetosella*—Sheep Sorrel.

Noctua augur.
Cratægus Oxyacantha—Whitethorn. After hybernation.
Salix Caprea—(Sallow). After hybernation.
Prunus spinosa—Blackthorn. After hybernation.
Also on various low plants.

Noctua plecta.
Galium verum—Lady's Bedstraw.
Asperula odorata—Sweet Woodruff.
Also on various low plants.

Noctua flammatra.
Taraxacum officinale—Dandelion.
Fragaria vesca—Wild Strawberry.

Noctua c-nigrum.
Salix repens—Dwarf Sallow.
Vaccinium Myrtillus—Whortleberry ; Bilberry.
Rumex—Dock. In confinement.
Plantago—Plantain. In confinement.
Lamium album—White Dead-nettle. In confinement.
Also on various low plants.

Noctua ditrapezium.
Polygonum aviculare—Knot-grass. (?)
Betula alba—Birch. In confinement.
Plantago—Plantain. In confinement.
Rumex—Dock. In confinement.
Salix—Sallow. In confinement.
Also on various low plants.

Noctua triangulum.
Salix repens—Dwarf Sallow.
Plantago—Plantain. In confinement.
Rumex—Dock. In confinement.
Salix—(Sallow). In confinement.
Cratægus Oxyacantha—Whitethorn. In confinement.
Rubus fruticosus—Bramble.
Lamium album—White Dead-nettle. In confinement.
Primula acaulis—Primrose.
Betula alba—Birch.
Also on various low plants.

Noctua stigmatica (rhomboidea).
Stellaria media—Chickweed.
Salix Caprea—(Sallow). After hybernation.
Also on other low plants.
In confinement on mint and carrot.

Noctua brunnea.
Salix cinerea—(Sallow).
„ *Caprea*—(Sallow).
Vaccinium Myrtillus—Whortleberry ; Bilberry.
Also on various low plants.

Noctua festiva.
Polygonum aviculare—Knot-grass. In confinement.
Salix cinerea—(Sallow).
„ *Caprea*—(Sallow).
Also on various low plants.

Noctua festiva v.conflua.
Silene acaulis—Moss Campion. Polyphagous.

Noctua dahlii.
Plantago—Plantain.
Rumex—Dock. In confinement, after hybernation.
Cratægus Oxyacantha—Whitethorn. In confinement, after hybernation.
Rubus fruticosus—Bramble. In confinement, after hybernation.
Also on various low plants.

Noctua subrosea.
Myrica Gale—Sweet Gale.
Salix alba—White Willow. In confinement.
„ *fragilis*—Crack Willow. In confinement.

Noctua rubi.
Salix repens—Dwarf Sallow.
Plantago—Plantain. In confinement.
Rumex—Dock.
Lamium album—White Dead-nettle. In confinement.
Also on grasses and various low plants.

Noctua umbrosa.
Scilla nutans—Squill; Blue-bell. On the seeds.
Poa annua—Annual Meadow-grass.
Rumex—Dock. In confinement.
Plantago lanceolata—Narrow-leaved Plantain. In confinement.
Galium Mollugo—Hedge Bedstraw. In confinement.
Also on other grasses and low plants.

Noctua baia (baja).
Salix repens—Dwarf Sallow.
Vaccinium Myrtillus—Whortleberry; Bilberry.
Plantago—Plantain. In confinement.
Rumex—Dock. In confinement.
Lamium album—White Dead-nettle. In confinement.
Also on various low plants.

Noctua sobrina.
Vaccinium Myrtillus—Whortleberry; Bilberry.
Cratægus Oxyacantha—Whitethorn. In confinement.

Noctua castanea (v. neglecta).
   Calluna Erica—Ling ; Heather.
   Salix cinerea—(Sallow).
     „ *Caprea*—(Sallow).

Noctua xanthographa.
   Poa annua—Annual Meadow-grass.
   Cratægus Oxyacantha—Whitethorn.
   Salix repens—Dwarf Sallow.
   Plantago—Plantain. In confinement.
   Rumex—Dock. In confinement.
   Lamium album—White Dead-nettle. In confinement.
   Also on other grasses.

Triphæna ianthina.
   Lamium purpureum—Purple Dead-nettle.
     „ *album*—White Dead-nettle.
   Primula acaulis—Primrose.
   Prunus spinosa—Blackthorn.
   Cratægus Oxyacantha—Whitethorn.
   Also on various garden and low plants.

Triphæna fimbria.
   Luzula maxima—Wood-rush.
   Lamium purpureum—Purple Dead-nettle.
     „ *album*—White Dead-nettle.
   Primula acaulis—Primrose.
   Also on other low plants.

Triphæna interjecta.
   Malva sylvestris—Mallow.
   Also on various low plants.

Triphæna orbona (subsequa).
   Agropyron (Triticum) repens—Couch Grass.
   Dactylis glomerata—Cock's-foot Grass.
   Potentilla reptans—Trailing Tormentil.
   Primula veris—Cowslip.
   Ranunculus veris—Meadow Crowfoot.
     „ *repens*—Creeping Crowfoot.

Triphæna comes (orbona).
   Salix cinerea—(Sallow). After hybernation.
     „ *Caprea*—(Sallow). After hybernation.
   Digitalis purpurea—Foxglove.
   Armeria maritima—Thrift.
   Cratægus Oxyacantha—Whitethorn. After hybernation.
   Stellaria media—Chickweed.
   Also on other low plants.

Triphæna pronuba.
  Stellaria media—Chickweed.
  Lychnis Githago—Corn-cockle.
  Also on very many low plants and vegetables.

Amphipyra pyramidea.
  Ulmus campestris—Elm.
  Quercus Robur—Oak.
  Salix alba—White Willow.

Amphipyra tragopogonis.
  Cratægus Oxyacantha—Whitethorn.
  Also on various low plants.

Mania typica.
  Prunus—Plum. Before hybernation.
  Epilobium hirsutum—Great Hairy Willow-herb.
  Rumex—Dock.
  Pyrus communis—Pear.
  Also on many other trees before hybernation.

Mania maura.
  Rumex—Dock.
  Stellaria media—Chickweed.
  Fragaria vesca—Wild Strawbbrry.

Panolis piniperda.
  Pinus sylvestris—Scotch Fir.

Pachnobia leucographa.
  Plantago—Plantain.

Pachnobia rubricosa.
  Rumex—Dock.
  Populus—Poplar.

Pachnobia hyperborea.
  (Food-plant at present unknown).

Tæniocampa gothica.
  Syringa vulgaris—Lilac.
  Salix Caprea—(Sallow).
  Quercus Robur—Oak.
  Rumex—Dock.
  Cratægus Oxyacantha—Whitethorn.
  Trifolium repens—Dutch Clover.
        ,,   *pratense*—Purple Clover.
  Also on laurel. Doubleday.
  Also on various trees and plants.

Tæniocampa incerta (instabilis).
  Salix Caprea—(Sallow).
  „ *alba*—White Willow.
  Quercus Robur—Oak.
  Rumex—Dock.
  Prunus spinosa—Blackthorn.
  Also on various trees and plants.

Tæniocampa opima.
  Salix Caprea—(Sallow.)
  Rosa spinosissima—Burnet Rose.
  Ammophila arundinacea—Marram.
  Senecio Jacobæa—Ragwort.
  Cynoglossum officinale—Hound's-tongue.

Tæniocampa populeti.
  Populus tremula—Aspen.
  Populus—Poplar.

Tæniocampa stabilis.
  Quercus Robur—Oak.
  Ulmus campestris—Elm.

Tæniocampa gracilis.
  Salix Caprea—(Sallow.)
  „ *alba*—White Willow.
  Spiræa Ulmaria—Meadow-sweet.

Tæniocampa miniosa.
  Quercus Robur—Oak.
  Cratægus Oxyacantha—Whitethorn.

Tæniocampa munda.
  Populus tremula—Aspen.
  Quercus Robur—Oak.
  Ulmus campestris—Elm.
  Prunus—Plum.

Tæniocampa pulverulenta (cruda).
  Quercus Robur—Oak.

Orthosia suspecta.
  (Food-plant at present unknown, probably on low plants.)

Orthosia upsilon.
  Salix alba—White Willow.
  „ *fragilis*—Crack Willow.
  Populus—Poplar.

Orthosia lota.
  Prunus spinosa—Blackthorn.
  Salix fragilis—Crack Willow.
  „ *Caprea*—(Sallow.)

Orthosia macilenta.
   Fagus sylvatica—Beech.

Anchocelis rufina.
   Quercus Robur—Oak.

Anchocelis pistacina.
   Ranunculus bulbosus—Bulbous Buttercup.
   Rumex—Dock.
   Also on other *Ranunculi*.

Anchocelis lunosa.
   Holcus lanatus—Meadow soft-grass (?).
   Also on other grasses.

Anchocelis litura.
   Spiræa Ulmaria—Meadow-sweet.
   Salix alba—White Willow.
   Rubus fruticosus—Bramble.

Cerastis (glæa) vaccinii.
   Ulmus campestris—Elm.
   Quercus Robur—Oak.
   Salix Caprea—(Sallow.)
   Also on low plants.

Cerastis (glæa) spadicea.
   Prunus spinosa—Blackthorn.
   Cratægus Oxyacantha—Whitethorn.
   Lonicera Periclymenum—Honeysuckle.
   Also on various low plants when about half-grown.

Cerastis (glæa) erythrocephala.
   Polygonum aviculare—Knot-grass (?).
   Also on various low plants.

Scopelosoma satellitia.
   Ulmus campestris—Elm.
   Quercus Robur—Oak.
   Fagus sylvatica—Beech.

Dasycampa rubiginea.
   Pyrus Malus—Apple.
   Taraxacum officinale—Dandelion.
   Quercus Robur—Oak.
   Prunus—Plum. In confinement.
   Polygonum aviculare—Knot-grass. In confinement.
   Also on low plants.

Oporina croceago.
   Quercus Robur—Oak.

Xanthia citrago.
   Tilia vulgaris—Lime.

Xanthia fulvago (cerago).
  Salix Caprea—(Sallow.) In the catkins.
  „ *cinera*—(Sallow.)
    Afterwards on low plants.

Xanthia flavago (silago).
  Salix Caprea—(Sallow).
  „ *cinerea*—(Sallow).
    Also on various low plants.

Xanthia aurago.
  Fagus sylvatica—Beech.
  Spiræa Filipendula—Dropwort.

Xanthia gilvago.
  Ulmus montana—Wych Elm.
  „ *campestris*—Elm.

Xanthia circellaris (ferruginea).
  Fraxinus excelsior—Ash.
  Ulmus montana—Wych Elm.
  Salix Caprea—(Sallow). On the buds.
  Populus tremula—Aspen. On the buds.

Cirrhædia xerampelina.
  Fraxinus excelsior—Ash.

Tethea subtusa.
  Populus—Poplar.

Tethea retusa.
  Populus—Poplar.
  Salix Caprea—(Sallow.)

Cosmia (Euperia) palacea (fulvago).
  Quercus Robur—Oak.
  Betula alba—Birch.

Dicycla oo.
  Quercus Robur—Oak.

Calymnia trapezina.
  Quercus Robur—Oak.
  Carpinus Betulus—Hornbeam.
  Betula alba—Birch.

Calymnia pyralina.
  Pyrus communis—Pear.
  Prunus—Plum.

Calymnia diffinis.
  Ulmus campestris—Elm.

Calymnia affinis.
  Ulmus campestris—Elm.

Eremobia (Ilarus) ochroleuca.
Aira præcox—Early Hair Grass (?).
Verbascum Thapsus—Mullein.
Dactylis glomerata—Cock's-foot Grass. In confinement, on the seeds.
Also on other grasses in dry places.

Dianthœcia luteago (v. barretti).
(Food-plant at present unknown.)

Dianthœcia cæsia.
Silene Cucubalus—White Campion. On the seeds.
„ *maritima*—Sea Campion. On the seeds.

Dianthœcia nana (conspersa).
Silene Cucubalus—White Campion. On the seeds.
„ *maritima*—Sea Campion. On the seeds.
„ *nutans*—Nottingham Catchfly. On the seeds.
Lychnis alba—Evening Campion. On the seeds.
„ *diurna*—Red Campion. On the seeds.
„ *Flos-cuculi*—Ragged Robin. On the seeds.

Dianthœcia albimacula.
Silene nutans—Nottingham Catchfly. On the seeds.
„ *Cucubalus*—White Campion. In confinement, on the seeds.
„ *maritima*—Sea Campion. In confinement, on the seeds.
Lychnis diurna—Red Campion. In confinement, on the seeds.

Dianthœcia compta.
(Doubtful British species.)

Dianthœcia capsincola.
Silene Cucubalus—White Campion. On the seeds.
„ *maritima*—Sea Campion. On the seeds.
Lychnis alba—Evening Campion. On the seeds.
„ *diurna*—Red Campion. On the seeds.

Dianthœcia cucubali.
Silene Cucubalus—White Campion. On the seeds.
„ *maritima*—Sea Campion. On the seeds.
Lychnis alba—Evening Campion. On the seeds.

Dianthœcia carpophaga.
Silene Cucubalus—White Campion. On the seeds.
Lychnis alba—Evening Campion. On the seeds.

Dianthœcia capsophila.
Silene Cucubalus—White Campion. In confinement, on the seeds.
„ *maritima*—Sea Campion. On the seeds.

Dianthœcia irregularis.
Silene Otites—Spanish Catchfly. On the flowers and seeds.
" *Cucubalus*—White Campion. In confinement.
Echium vulgare—Viper's Bugloss.

Hecatera chrysozona (dysodea).
Lactuca virosa—Sleepwort. On the flowers and seeds.
Also on other *Lactucæ*.

Hecatera serena.
Lactuca virosa—Sleepwort.
Sonchus oleraceus—Sow-thistle.
" *arvensis*—Corn Sow-thistle.
Hieracium Pilosella—Mouse-ear Hawk-weed.
Also on other *Hieracia*.
In confinement on lettuce.

Polia chi.
Lactuca virosa—Sleepwort (?).
Sonchus oleraceus—Sow-thistle.
" *arvensis*—Corn Sow-thistle.
Cratægus Oxyacantha—Whitethorn.
Rumex crispus—Curled Dock. In confinement.
Salix Caprea—Sallow.
" *cinerea*—Sallow. In confinement.
" *viminalis*—Osier. In confinement.

Polia flavicincta.
Rumex—Dock.
Plantago—Plantain.
Digitalis purpurea—Foxglove.
Senecio vulgaris—Groundsel.
Lonicera Periclymenum—Honeysuckle.
Hedera Helix—Ivy.
Stellaria media—Chickweed.
Lathyrus (Orobus) sylvestris—Everlasting Pea.

Polia xanthomista.
Armeria maritima—Thrift.
Campanula rotundifolia—Harebell.
Silene gallica—English Catchfly.
" *acaulis*—Moss Campion.
" *nutans*—Nottingham Catchfly.
" *Cucubalus*—White Campion.
" *maritima*—Sea Campion.
Plantago maritima—Sea Plantain. In confinement.
Viola canina—Dog Violet.
" *tricolor*—Heartsease.
" *odorata*—Sweet Violet.

Dasypolia templi.
Heracleum Sphondylium—Cow-parsnep.

Epunda lichenea.
Senecio Jacobœa—Ragwort.
Also on *Sedum*.

Epunda lutulenta.
Lithospermum arvense—Corn Gromwell; Bastard Alkaret.
Scabiosa arvensis—Field Scabious.
Calluna Erica—Ling; Heather.
Mentha—Mint.
Poa annua—Annual Meadow-grass. In confinement.
Potentilla Fragariastrum—Strawberry-leaved Cinquefoil. In confinement.

Epunda nigra.
Rumex—Dock.
Armeria maritima—Thrift.
Galium Mollugo—Hedge Bedstraw.
Stellaria media—Chickweed.

Cleoceris viminalis.
Salix cinerea—(Sallow.)
  „ *alba*—White Willow (?).

Miselia oxyacanthæ.
Cratægus Oxyacantha—Whitethorn.
Prunus spinosa—Blackthorn.

Miselia bimaculosa.
Ulmus campestris—Elm.

Agriopis aprilina.
Quercus Robur—Oak.

Valeria oleagina.
Prunus spinosa—Blackthorn.

Euplexia lucipara.
Digitalis purpurea—Foxglove.
Pteris aquilina—Brake; Bracken.
Lastrœa (Nephrodium) Filix-mas—Male Fern.

Phlogophora meticulosa.
Senecio vulgaris—Groundsel.
Also on many other low plants.

Trigonophora flammea (empyrea).
Ranunculus bulbosus—Bulbous Buttercup.
  „ *Ficaria*—Pilewort.
Also on various low plants.

Aplecta prasina (herbida).
Polygonum aviculare—Knot-grass.
Rumex—Dock.
Also on other low plants.

Aplecta occulta.
Taraxacum officinale—Dandelion.
Primula acaulis—Primrose.
Polygonum aviculare—Knot-grass. In confinement.
Calluna Erica—Ling ; Heather.
Salix—(Sallow.) In confinement.
Rubus fruticosus—Bramble. In confinement.
Betula alba—Birch. In confinement.
Rumex pulcher—Fiddle Dock. In confinement.
Vinca major—Greater Periwinkle. In confinement.
Plantago lanceolata—Narrow-leaved Plantain. In confinement.
Also on various low plants.

Aplecta nebulosa.
Cratægus Oxyacantha—Whitethorn.
Rumex—Dock.
Betula alba—Birch.
Salix Caprea—(Sallow.)
On various low plants before, and on other trees after, hybernation.

Aplecta tincta.
Betula alba—Birch.
Poa annua—Annual Meadow-grass.
Also on other grasses and low plants.

Aplecta advena.
Polygonum aviculare—Knot-grass.
Also on lettuce and other low plants.

Crymodes exulis.
Poa annua—Annual Meadow-grass. On this and other species of *Poa*.

Hadena porphyrea (satura).
Lonicera Periclymenum—Honeysuckle.
Humulus Lupulus—Hop.

Hadena adusta.
Salix Caprea—(Sallow)
Also on various low plants.

Hadena protea.
Quercus Robur—Oak.

Hadena glauca.
Salix Caprea—(Sallow.) In confinement on lettuce.

Hadena dentina.
   Taraxacum officinale—Dandelion.

Hadena trifolii (chenopodii).
   Chenopodium Vulvaria—Stinking Goose-foot ; Orach.
      „   *album*—White Goose-foot.
      „   *Bonus-Henricus*—Good King Henry.
   Polygonum aviculare—Knot-grsss.

Hadena atriplicis.
   Chenopodium vulvaria—Stinking Goose-foot ; Orach.
      „   *album*—White Goose-foot.
      „   *Bonus-Henricus*—Good King Henry.
   Atriplex patula—Orach.
      „   *laciniata*—Sea Orach.
   Polygonum aviculare—Knot-grass.   Also on other *Polygona*.
   Rumex—Dock.
   Also on various low plants.

Hadena dissimilis (suasa).
   Polygonum aviculare—Knot-grass
   Plantago major—Greater Plantain.   In confinement.
   Also on other low plants.

Hadena oleracea.
   Rumex—Dock.
   Urtica dioica—Stinging Nettle.
   Also on various low plants.

Hadena pisi.
   Cytisus (Sarothamnus) scoparius—Broom.
   Pteris aquilina—Brake ; Bracken.
   Also on other plants and shrubs.

Hadena thalassina.
   Cytisus (Sarothamnus) scoparius—Broom.
   Lonicera Periclymenum—Honeysuckle.
   Polygonum aviculare—Knot-grass.

*****Hadena peregrina.**
   Chenopodium Vulvaria—Stinking Goose-foot ; Orach.
      „   *album*—White Goose-foot.
      „   *Bonus-Henricus*—Good King Henry.
   Salsola Kali—Prickly Glass-wort.

Hadena genistæ.
   Stellaria media—Chickweed.   In confinement.
   Polygonum aviculare—Knot-grass.   Also on other *Polygona*.

Hadena rectilinea.
   Impatiens Noli-tangere—Do-not-touch-me.
   Rubus fruticosus—Bramble.

Hadena rectilinea—*continued.*
   Vaccinium Vitis-Idæa—Cow-berry.
      ,,   *Myrtillus*—Whortleberry; Bilberry.
   Salix Caprea—(Sallow.)

Hadena contigua.
   Myrica Gale—Sweet Gale.
   Betula alba—Birch.
   Corylus Avellana—Hazel.
   Quercus Robur—Oak.
   Salix repens—Dwarf Sallow.

Xylocampa areola (lithoriza).
   Lonicera Periclymenum—Honeysuckle.

Xylomiges conspicillaris.
   Prunus spinosa—Blackthorn.
   Lotus corniculatus—Bird's-foot Trefoil.
   Polygonum aviculare—Knot-grass.

Calocampa vetusta.
   Iris Pseudacorus—Yellow Iris.
   Rumex—Dock.
   Scabiosa succisa—Devil's-bit Scabious.
      ,,   *arvensis*—Field Scabious.
      ,,   *Columbaria*—Small Scabious.
   Trifolium procumbens—Hop Trefoil.
      ,,   *ochroleucon.*
   Also on various meadow plants.

Calocampa exoleta.
   Cnicus arvensis—Field Thistle.
   Scabiosa succisa—Devil's-bit Scabious.
   Ononis spinosa—Rest-harrow.
   Also on various meadow plants.

Calocampa solidaginis.
   Vaccinium Myrtillus—Whortleberry; Bilberry.
   Cratægus Oxyacantha—Whitethorn.

Xylina ornithopus (rhizolitha).
   Quercus Robur—Oak.

Xylina semibrunnea.
   Fraxinus excelsior—Ash.
   Salix alba—White Willow.
   Quercus Robur—Oak.

Xylina socia (petrificata).
   Quercus Robur—Oak.
   Tilia vulgaris—Lime.
   Also on other trees.

LARVÆ AND FOOD-PLANTS.

Xylina furcifera (conformis).
  Alnus glutinosa—Alder.
  Betula alba—Birch.

Xylina lambda.
  Alnus glutinosa—Alder.
  Myrica Gale—Sweet Gale.

Asteroscopus nubeculosa.
  Quercus Robur—Oak.
  Betula alba—Birch.

Asteroscopus sphinx (cassinea).
  Quercus Robur—Oak.
  Tilia vulgaris—Lime.
  Salix cinerea—(Sallow.)
  „ *Caprea*—(Sallow.)
  Also on other trees.

Cucullia verbasci.
  Verbascum Thapsus—Mullein.
  „ *nigrum*—Black Mullein.
  Scrophularia aquatica—Water Fig-wort.
  „ *nodosa*—Knotted Fig-wort.

Cucullia scrophulariæ.
  Verbascum Blattaria—Moth Mullein.
  Scrophularia aquatica—Water Fig-wort.
  „ *nodosa*—Knotted Fig-wort.

Cucullia lychnitis.
  Verbascum Lychnitis—White Mullein.
  „ *nigrum*—Black Mullein.

Cucullia asteris.
  Solidago Virgaurea—Golden-rod.
  Aster Tripolium—Sea Star-wort.
  Also in gardens on China Asters.

Cucullia gnaphalii.
  Solidago Virgaurea—Golden-rod.

Cucullia absinthii.
  Artemisia Absinthium—Common Wormwood.

Cucullia chamomillæ.
  Anthemis Cotula.
  „ *nobilis*—Chamomile.
  „ *arvensis*—Corn Chamomile.
  Chrysanthemum Parthenium—Common Feverfew.
  Matricaria inodora—Corn Feverfew.
  „ *Chamomilla*—Dog's Chamomile.

E

* **Cucullia artemisiæ.**
    Artemisia Absinthium—Common Wormwood.
* **Cucullia umbratica.**
    Lactuca virosa—Sleepwort.
    Sonchus oleraceus—Sow-thistle.
      ,,   *arvensis*—Corn Sow-thistle.
      ,,   *palustris*—Marsh Sow-thistle.
* **Gonoptera libatrix.**
    Salix Caprea—(Sallow).
      ,,   *alba*—White Willow.
* **Habrostola tripartita (urticæ).**
    Urtica dioica—Stinging Nettle.
* **Habrostola triplasia.**
    Humulus Lupulus—Hop.
    Urtica dioica—Stinging Nettle.
* **Plusia chryson (orichalcea).**
    Eupatorium cannabinum—Hemp Agrimony.
* **Plusia chrysitis.**
    Arctium minus (Lappa)—Burdock.
    Lamium album—White Dead-nettle.
    Urtica dioica—Stinging Nettle.
* **Plusia bractea.**
    Anthriscus sylvestris—Hare's Parsley.
    Eupatorium cannabinum—Hemp Agrimony.
    Senecio vulgaris—Groundsel.
    Hieracium Pilosella—Mouse-ear Hawk-weed.
    Mercurialis perennis—Dog's Mercury.
    Lamium purpureum—Purple Dead-nettle.
      ,,   *album*—White Dead-nettle.
* **Plusia festucæ.**
    Iris Pseudacorus—Yellow Iris.
    Sparganium simplex—Unbranched Bur-reed.
    Carex—Sedge.
* **Plusia iota.**
    Lonicera Periclymenum—Honeysuckle.
    Senecio vulgaris—Groundsel.
    Lamium album—White Dead-nettle.
    Urtica dioica—Stinging Nettle.
    Also on various low plants.
* **Plusia pulchrina.**
    Senecio vulgaris—Groundsel.
    Urtica dioica—Stinging Nettle.
      ,,   *urens*—Small Nettle.
    Also on various low plants.

Plusia gamma.
Humulus Lupulus—Hop.
Also on all sorts of low plants.

Plusia interrogationis.
Urtica dioica—Stinging Nettle.

Plusia ni.
Polygonum aviculare—Knot-grass (?).
Also on other low plants.

Anarta melanopa.
Arctostaphylos Uva-ursi—Red Bear-berry.
Phyllodoce (menziesia) taxifolia—Scottish Menziesia.

Anarta cordigera.
Vaccinium Vitis-Idæa—Cow-berry.
„ *Myrtillus*—Whortleberry; Bilberry.
„ *uliginosum*—Bog Whortleberry.

Anarta myrtilli.
Calluna Erica—Ling; Heather.
Salix alba—White Willow. In confinement.

Heliaca (Heliodes) tenebrata (arbuti).
Cerastium glomeratum—Mouse-ear Chickweed.
„ *arvense*—Field Mouse-ear Chickweed.

Heliothis dipsacea.
Linaria vulgaris—Yellow Toad-flax.
Silene Otites—Spanish Catchfly.
Lychnis alba—Evening Campion.
Ononis spinosa—Rest-harrow.
Crepis virens—Smooth Hawk's-beard.
Also on various low plants and other *Silene*.

Heliothis scutosa.
Artemisia vulgaris—Mugwort.
„ *campestris*—Field Wormwood.

Heliothis peltigera.
Hyoscyamus niger—Henbane.
Matricaria inodora—Corn Feverfew.
Ononis spinosa—Rest-harrow.
Lepigonum (Spergularia) rubrum—Purple Sandwort.
Arenaria—Sandwort.

Heliothis armigera.
Reseda lutea—Wild Mignonette.
Also on scarlet-geranium. Ent. ix. 261.

Chariclea umbra (marginata).
Polygonum aviculare—Knot-grass. In confinement.
Ononis spinosa—Rest-harrow.

Agrophila trabealis (sulphuralis).
   Convolvulus arvensis—Field Bindweed.

Acontia luctuosa.
   Convolvulus arvensis—Field Bindweed.

* Acontia solaris.
   Polygonum aviculare—Knot-grass (?).
   Also on other low plants.

Erastria venustula.
   Rumex Acetosella—Sheep Sorrel.
   Rubus fruticosus—Bramble.
   Potentilla reptans—Trailing Tormentil. On the flowers.
      „   *Anserina*—Silver-weed. On the flowers.

Erastria fasciana (fuscula).
   Rubus Idæus—Raspberry.
      „ *fruticosus*—Bramble.
   Molinia cærulea—Purple Melic-grass.

Bankia argentula.
   Poa annua—Annual Meadow-grass.
      „ *pratensis.*
   Glyceria aquatica.
   Also on other grasses.

Hydrelia uncula (unca).
   Carex—Sedge.
   Carex sylvatica. In confinement.

Thalpochares (Micra) ostrina.
   Carduus—Thistle (?).
   Cnicus—Thistle (?).

Thalpochares (Micra) parva.
   (Food-plant unknown; abroad on the seeds of *Inula montana.*)

Thalpochares (Micra) paula.
   (Food-plant unknown; abroad on *Gnaphalium arenarium*).

Phytometra viridaria (ænea).
   Polygala vulgaris—Common Milk-wort.

Euclidia mi.
   Melilotus arvensis—Melilot.
   Trifolium pratense—Purple Clover.
      „ *repens*—Dutch Clover.

Euclidia glyphica.
   Trifolium repens—Dutch Clover.

Ophiodes lunaris.
   Quercus Robur—Oak.

Catephia alchymista.
Quercus Robur—Oak.

Catocala fraxini.
Fraxinus excelsior—Ash.
Populus—Poplar.
  „   *tremula*—Aspen.

Catocala nupta.
Prunus—Plum.
Salix alba—(Sallow).
  „   *fragilis*—Crack Willow.
Populus—Poplar.

Catocala promissa.
Quercus Robur—Oak.

Catocala sponsa.
Quercus Robur—Oak.

Aventia flexula.
Lichen. On thorns, cherry and yew.

Toxocampa pastinum.
Vicia Cracca—Tufted Vetch.

Toxocampa craccæ.
Vicia sylvatica—Wood Vetch.

Boletobia fuliginaria.
Fungi. In damp cellars, etc.

Rivula sericealis.
Brachypodium sylvaticum.
  „   *pinnatum.*
Glyceria maritima

Zanclognatha grisealis.
Chrysosplenium alternifolium—Alternate-leaved Golden Saxifrage.

Zanclognatha tarsipennalis.
Rubus Idæus—Raspberry.

Zanclognatha emortualis.
Quercus Robur—Oak. Dead leaves preferred.

Herminia cribralis.
Carex sylvatica.
Luzula vernalis (pilosa)—Broad-leaved Wood-rush.
Salix—Sallow.

Herminia derivalis.
Castanea sativa—Chestnut. In confinement on the dead leaves.
Quercus Robur—Oak. In confinement on the dead leaves; also on grass.

Pechypogon barbalis.
   Quercus Robur—Oak.
   Betula alba—Birch.

Madopa salicalis.
   Salix Caprea—(Sallow).
   „ *cinerea*—(Sallow).
   „ *alba*—White Willow.

Bomolocha fontis (crassalis).
   Erica cinerea—Common Purple Heath.
   „ *Tetralix*—Cross-leaved Heath.

Hypena rostralis.
   Humulus Lupulus—Hop.

Hypena proboscidalis.
   Urtica dioica—Stinging Nettle.

Hypenodes albistrigalis.
   (Food-plant at present unknown).

Hypenodes costæstrigalis.
   Thymus Serpyllum—Thyme.   In confinement.

Tholomiges (schrankia) turfosalis.
   (Food-plant at present unknown).

Brephos parthenias.
   Fagus sylvatica—Beech.
   Betula alba—Birch.
   Quercus Robur—Oak.

Brephos notha.
   Salix cinerea—(Sallow).
   Populus tremula—Aspen.

GEOMETRÆ.

Uropteryx sambucaria.
   Prunus spinosa—Blackthorn.
   Rubus fruticosus—Bramble.
   Cratægus Oxyacantha—Whitethorn.
   Hedera Helix—Ivy.
   Sambucus nigra—Elder.
   Lonicera Periclymenum—Honeysuckle.
   Myosotis arvensis—Field Forget-me-not.
   Quercus Robur—Oak.
   Also on various herbaceous plants.

Epione paralellaria (vespertaria).
   Corylus Avellana—Hazel.

Epione apiciaria.
Alnus glutinosa—Alder.
Corylus Avellana—Hazel.
Salix alba—White Willow.
Populus—Poplar.

Epione advenaria.
Vaccinium Myrtillus—Whortleberry; Bilberry.

Rumia luteolata (cratægata).
Prunus spinosa—Blackthorn.
Pyrus Malus—Apple.
Cratægus Oxyacantha—Whitethorn.

Venilia macularia (maculata).
Lamium purpureum—Purple Dead-nettle.
Urtica dioica—Stinging Nettle.
Also on various herbaceous plants.

Angerona prunaria.
Clematis vitalba—Traveller's Joy.
Cytisus (Sarothamnus) scoparius—Broom.
Prunus spinosa—Blackthorn.
    „   —Plum.
Fagus sylvatica—Beech.

Metrocampa margaritaria.
Ulmus campestris—Elm.
Betula alba—Birch.
Carpinus Betulus—Hornbeam.
Quercus Robur—Oak.
Fagus sylvatica—Beech.

Ellopia prosapiaria (fasciaria).
Pinus sylvestris—Scotch Fir.

Eurymene dolobraria.
Betula alba—Birch.
Quercus Robur—Oak.
Fagus sylvatica—Beech.

Pericallia syringaria.
Sambucus nigra—Elder.
Ligustrum vulgare—Privet.
Syringa vulgaris—Lilac.

Selenia bilunaria (illunaria).
Prunus—Plum.
Salix alba—White Willow.

Selenia lunaria.
Prunus spinosa—Blackthorn.
Ulmus campestris—Elm.
Quercus Robur—Oak.

Selenia tetralunaria (illustraria).
   Fraxinus excelsior—Ash.
   Betula alba—Birch.
   Quercus Robur—Oak.
   Fagus sylvatica—Beech.

Odontopera bidentata.
   Quercus Robur—Oak.

Crocallis elinguaria.
   Fagus sylvatica—Beech.
   Lonicera Periclymenum—Honeysuckle.
   Cratægus Oxyacantha—Whitethorn.
   Pyrus Malus—Apple.
   Prunus spinosa—Blackthorn.
   Also on other trees.

Eugonia (Ennomos) autumnaria (alniaria).
   Fagus sylvatica—Beech.
   Quercus Robur—Oak.
   Alnus glutinosa—Alder.
   Betula alba—Birch.
   Sambucus nigra—Elder.
   Pyrus Malus—Apple.
   Acer Pseudo-platanus—Sycamore.
   Cratægus Oxyacantha—Whitethorn.   In confinement.
   Prunus—Plum.   In confinement.
   Salix—(Sallow).   In confinement.
   Also on apricot.

Eugonia (Ennomos) alniaria (tiliaria).
   Quercus Robur—Oak.
   Betula alba—Birch.
   Salix—(Sallow).   In confinement.

Eugonia (Ennomos) fuscantaria.
   Fraxinus excelsior—Ash.
   Ligustrum vulgare—Privet.

Eugonia (Ennomos) erosaria.
   Quercus Robur—Oak.
   Betula alba—Birch.

Eugonia (Ennomos) quercinaria (angularia)
   Syringa vulgaris—Lilac.
   Cratægus Oxyacantha—Whitethorn.
   Quercus Robur—Oak.
   Betula alba—Birch.
   Ulmus campestris—Elm.

Himera pennaria.
   Quercus Robur—Oak.

Phigalia pedaria (pilosaria).
  Quercus Robur—Oak.

Nyssia zonaria.
  Lotus corniculatus—Bird's-foot Trefoil.
  Salix Caprea—(Sallow).
  „ *cinerea*—(Sallow).
  Centaurea nigra—Knap-weed.
  Achillea Millefolium—Yarrow; Milfoil.

Nyssia hispidaria.
  Quercus Robur—Oak.

Nyssia lapponaria.
  Pyrus Malus—Apple.

Biston hirtaria.
  Ulmus campestris—Elm.
  Prunus—Plum.
  Tilia vulgaris—Lime.
  Also on fruit and other trees.

Amphidasys strataria (prodomaria).
  Quercus Robur—Oak.
  Betula alba—Birch.

Amphidasys betularia.
  Robinia Pseud-acacia—Acacia.
  Populus—Poplar.
  Salix alba—White Willow.
  Fagus sylvatica—Beech.
  Quercus Robur—Oak.
  Alnus glutinosa—Alder.
  Betula alba—Birch.
  Ulmus campestris—Elm.
  Pyrus Malus—Apple.
  Acer Pseudo-platanus—Sycamore.
  Tilia vulgaris—Lime.

Hemerophila abruptaria.
  Rosa canina—Dog-rose (?).
  Ligustrum vulgare—Privet.
  Also on cultivated rose.

Cleora angularia (viduaria).
  Lichen. On trees.

Cleora glabraria.
  Lichen. On firs, and oaks.

Cleora lichenaria.
  Lichen. On trees.

Boarmia repandata.
Euonymus europæus—Spindle.
Prunus spinosa—Blackthorn.
„   —Plum.
Rubus fruticosus—Bramble.
Cratægus Oxyacantha—Whitethorn.
Vaccinium Myrtillus—Whortleberry ; Bilberry.
Calluna Erica—Ling ; Heather.
Betula alba—Birch.

Boarmia gemmaria (rhomboidaria).
Prunus—Plum.
Rosa canina—Dog Rose (?).
Hedera Helix—Ivy.
Betula alba—Birch.
Quercus Robur—Oak.
Also on cultivated rose.

Boarmia abietaria.
Betula alba—Birch.
Pinus—Fir.
Syringa vulgaris—Lilac.

Boarmia cinctaria.
Erica Tetralix—Cross-leaved Heath.
„   *cinerea*—Common Purple Heath.

Boarmia roboraria.
Quercus Robur—Oak.

Boarmia consortaria.
Quercus Robur—Oak.

Tephrosia consonaria.
Fagus sylvatica—Beech.
Carpinus Betulus—Hornbeam.
Betula alba—Birch.
Quercus Robur—Oak.

Tephrosia crepuscularia.
Pinus—Fir.
Populus—Poplar.
Salix alba—White Willow.
Quercus Robur—Oak.
Alnus glutinosa—Alder.
Betula alba—Birch.
Ulmus campestris—Elm.
Prunus spinosa—Blackthorn.

Tephrosia biundularia.
Larix europæa—Larch.
Pinus—Fir.

Tephrosia biundularia—*continued.*
Quercus Robur—Oak.
Alnus glutinosa—Alder.
Betula alba—Birch.
Prunus—Plum. In confinement.

Tephrosia luridata (extersaria).
Betula alba—Birch.

Tephrosia punctularia.
Betula alba—Birch.

Gnophos obscuraria.
Poterium Sanguisorba—Common Salad Burnet
Fragaria vesca—Wild Strawberry.
Helianthemum Chamæcistus—Rock Rose.

Dasydia obfuscaria.
Vicia sativa—Common Vetch.
Genista tinctoria—Dyer's Green-weed.
Calluna Erica—Ling; Heather.
Polygonum aviculare—Knot-grass. In confinement.

Psodos coracina (trepidaria).
Mosses.

* **Tephrona sepiaria.**
Lichen. On walls.

Pseudoterpna pruinata (cytisaria).
Genista anglica—Needle Furze.
Cytisus (Sarothamnus) scoparius—Broom
Ulex europæus—Furze; Gorse.

Geometra papilionaria.
Betula alba—Birch.
Corylus Avellana—Hazel.
Fagus sylvatica.—Beech.

Geometra vernaria.
Clematis Vitalba—Traveller's Joy.

Phorodesma pustulata.
Quercus Robur—Oak.

Phorodesma smaragdaria.
Artemisia maritima—Sea Wormwood.

Nemoria viridata.
Rubus fruticosus—Bramble.
Cratægus Oxyacantha—Whitethorn.

Iodis lactearia.
Betula alba—Birch.

Hemithea strigata (thymiaria).
Cratægus Oxyacantha—Whitethorn.
Quercus Robur—Oak.

Zonosoma (Ephyra) porata.
Quercus Robur—Oak.

Zonosoma (Ephyra) punctaria.
Quercus Robur—Oak.

Zonosoma (Ephyra) linearia (trilinearia).
Fagus sylvatica—Beech.

Zonosoma (Ephyra) annulata (omicronaria).
Acer campestre—Maple.
„ *Pseudo-platanus*—Sycamore.

Zonosoma (Ephyra) orbicularia.
Alnus glutinosa—Alder.
Salix cinerea—(Sallow).
„ *Caprea*—(Sallow).

Zonosoma (Ephyra) pendularia.
Betula alba—Birch.

Hyria muricata (auroraria).
Plantago—Plantain.
Polygonum aviculare—Knot-grass.

Asthena luteata.
Alnus glutinosa—Alder.
Acer campestre—Maple.

Asthena candidata.
Carpinus Betulus—Hornbeam.
Corylus Avellana—Hazel.

Asthena sylvata.
Betula alba—Birch.
Alnus glutinosa—Alder.
Prunus spinosa—Blackthorn.

Asthena blomeri (blomeraria).
Ulmus montana—Wych Elm.

Eupisteria obliterata (heperata).
Alnus glutinosa—Alder.

Venusia cambrica (cambricaria).
Pyrus Aucuparia—Mountain Ash.

Acidalia perochraria (ochrearia).
Silene maritima—Sea Campion.
Galium verum—Lady's Bedstraw. On the flowers. In confinement.
„ *palustre*—Water Bedstraw. On the flowers. In confinement.

Acidalia ochrata.
Ononis spinosa—Rest-harrow.
Galium verum—Lady's Bedstraw. On the flowers.
Solidago Virgaurea—Golden-rod.
Tussilago Farfara—Colt's-foot. On the flowers.
Crepis virens—Smooth Hawk's-beard. On the flowers.
Taraxacum officinale—Dandelion.
Festuca ovina—Sheep's Fescue Grass (?).

Acidalia rubiginata (rubricata).
Lotus corniculatus—Bird's-foot Trefoil. In confinement.
 ,, *pilosus*—Greater Bird's-foot Trefoil. In confinement on this and various *Trifolii*.
Thymus Serpyllum—Thyme.
Prunus—Plum.
Salix—Sallow.

Acidalia dimidiata (scutulata).
Pimpinella Saxifraga—Common Burnet Saxifrage. On the flowers.
Anthriscus sylvestris—Hare's Parsley. On the flowers.
Galium Mollugo—Hedge Bedstraw.
Taraxacum officinale—Dandelion. In confinement.

Acidalia bisetata.
Taraxacum officinale—Dandelion.
Polygonum aviculare—Knot-grass. In confinement.
Rubus fruticosus—Bramble. In confinement; on the withered leaves.

Acidalia trigeminata.
Polygonum aviculare—Knot-grass. In confinement.
Cratægus Oxyacantha—Whitethorn. In confinement.
Stellaria media—Chickweed. In confinement.

Acidalia contiguaria.
Erica Tetralix—Cross-leaved Heath.
 ,, *cinerea*—Common Purple Heath.
Polygonum aviculare—Knot-grass.
Empetrum nigrum—Black Crow-berry.
Stellaria media—Chickweed.
Calluna Erica—Ling; Heather. In confinement.
Cratægus Oxyacantha—Whitethorn. In confinement.

Acidalia rusticata.
Parietaria officinalis—Pellitory-of-the-wall.
Hedera Helix—Ivy. In confinement.
Syringa vulgaris—Lilac. In confinement.
Rubus fruticosus—Bramble. In confinement, after hybernation, on withered leaves.

Acidalia holosericata.
Polygonum aviculare—Knot-grass. In confinement.
Helianthemum Chamæcistus—Rock Rose. In confinement.

Acidalia dilutaria (osseata).
Polygonum aviculare—Knot-grass. In confinement.
Taraxacum officinale—Dandelion. In confinement.
Anagallis arvensis—Scarlet Pimpernel. In confinement.
Also on moss.

Acidalia virgularia.
Polygonum aviculare—Knot-grass. In confinement.
Taraxacum officinale—Dandelion. In confinement.

Acidalia circellata.
Polygonum aviculare. Knot-grass. In confinement.

Acidalia ornata.
Thymus Serpyllum—Thyme.
Origanum vulgare—Marjoram.
Mentha—Mint. In confinement.

Acidalia marginepunctata (promutata).
Dianthus Armeria—Deptford Pink.
Vicia sativa—Common Vetch.
Spiræa Ulmaria—Meadow-sweet.
Valeriana officinalis—Great Wild Valerian.
Achillea Millefolium—Yarrow ; Milfoil.
Plantago—Plantain.
Nepeta Glechoma—Ground Ivy.

Acidalia straminata.
Polygonum aviculare—Knot-grass. In confinement.
Thymus Serpyllum—Thyme.
Chenopodium album—White Goose-foot.

Acidalia subsericeata.
Polygonum aviculare—Knot-grass. In confinement.

Acidalia immutata.
Stellaria media—Chickweed. In confinement.
Senecio vulgaris—Groundsel. In confinement.
Taraxacum officinale—Dandelion. In confinement.
Cratægus Oxyacantha—Whitethorn. In confinement, after hybernation.
Clematis Vitalba—Traveller's joy. In confinement, after hybernation.
Lythrum Salicaria—Purple Loosestrife.
Plantago—Plantain.
Achillea Millefolium—Yarrow ; Milfoil.
Valeriana officinalis—Great Wild Valerian.
Spiræa Ulmaria—Meadow-sweet;

Acidalia immorata.
  Calluna Erica—Ling ; Heather.
Acidalia strigaria.
  (Doubtful British species ; food-plant unknown).
Acidalia remutaria.
  Polygonum aviculare—Knot-grass. In confinement.
  Rumex—Dock. On the withered leaves in confinement.
Acidalia fumata.
  Erica Tetralix—Cross-leaved Heath.
    ,,   *cinerea*—Common Purple Heath.
Acidalia strigilaria.
  Clematis Vitalba—Traveller's Joy.
  Stachys sylvatica—Hedge Wound-wort.
  Polygonum aviculare. Knot-grass. In confinement.
  Taraxacum officinale—Dandelion. In confinement.
Acidalia imitaria
  Rumex Acetosella—Sheep Sorrel. In confinement.
  Cytisus (Sarothamnus) scoparius—Broom.
Acidalia emutaria.
  Polygonum aviculare—Knot-grass. In confinement.
  Silene maritima—Sea Campion. In confinement.
  Lotus corniculatus—Bird's-foot Trefoil. In confinement.
    ,, *pilosus*—Greater Bird's-foot Trefoil. In confinement.
  Stellaria media—Chickweed. In confinement.
Acidalia aversata.
  Spiræa Ulmaria—Meadow-sweet.
  Geum urbanum—Common Avens.
    ,,   *rivale*—Water Avens.
  Polygonum aviculare—Knot-grass.
  Also on other plants in hedges.
Acidalia inornata.
  Salix alba—White Willow. On the low shoots and also on various low plants.
  Polygonum aviculare—Knot-grass. In confinement.
Acidalia degenaria.
  Rubus fruticosus—Bramble. In confinement.
  Stellaria media—Chickweed. In confinement.
  Polygonum aviculare—Knot-grass. In confinement.
  Cratægus Oxyacantha—Whitethorn. In confinement.
  Lonicera Periclymenum—Honeysuckle. In confinement.
  Potentilla. In confinement.
Acidalia emarginata.
  Galium verum—Lady's Bedstraw.
  Convolvulus arvensis—Field Bindweed.

* Acidalia herbariata.
    Dried Plants.
Timandra amataria.
    Rumex—Dock.
    „   *acetosa*—Sorrel.
    Polygonum aviculare—Knot-grass.
Cabera pusaria.
    Quercus Robur—Oak.
    Corylus Avellana—Hazel.
    Alnus glutinosa—Alder.
    Betula alba—Birch.
Cabera rotundaria.
    Alnus glutinosa—Alder (?).
    Betula alba—Birch.
Cabera exanthemaria.
    Salix Caprea—(Sallow).
    „   *cinerea*—(Sallow).
    Alnus glutinosa—Alder.
Bapta (Corycia) temerata.
    Prunus Avium—Wild Cherry.
    „   *spinosa*—Blackthorn.
Bapta (Corycia) bimaculata (taminata).
    Prunus Avium—Wild Cherry.
Aleucis pictaria.
    Prunus spinosa—Blackthorn.
Macaria alternata.
    Salix Caprea—(Sallow).
    Alnus glutinosa—Alder.
Macaria notata.
    Salix Caprea—(Sallow).
    „   *cinerea*—Sallow).
Macaria liturata.
    Pinus—Fir.
Halia vauaria (wavaria).
    Ribes Grossularia—Gooseberry.
    „   *rubrum*—Red Currant.
Halia brunneata (pinetaria).
    Vaccinium Myrtillus—Whortleberry; Bilberry.
Strenia clathrata.
    Onobrychis viciæfolia (sativa)—Saint-foin.
    Trifolium repens—Dutch Clover.
    Medicago sativa—Lucerne.
    Also on other leguminous plants.

Penagra petraria.
Pteris aquilina—Brake ; Bracken.

Numeria pulveraria.
Salix Caprea—(Sallow).
  „ *cinerea*—(Sallow).

Scodiona belgiaria.
Salix alba—White Willow.
Erica cinerea—Common Purple Heath.
  „ *Tetralix*—Cross-leaved Heath.
Calluna Erica—Ling ; Heather.

Selidosema ericetaria (plumaria).
Calluna Erica—Ling ; Heather.
Erica cinerea—Common Purple Heath.
  „ *Tetralix*—Cross-leaved Heath.
Lotus corniculatus—Bird's-foot Trefoil.

Fidonia carbonaria.
Salix Caprea—(Sallow).
  „ *cinerea*—(Sallow).
Betula alba—Birch.

Fidonia limbaria (conspicuata)
Cytisus (Sarothamnus) scoparius—Broom.

Ematurga atomaria.
Lotus corniculatus—Bird's-foot Trefoil.
  „ *pilosus*—Greater Bird's-foot Trefoil.
Centaurea nigra—Knapweed.
Also on various plants on heaths.

Bupalus piniaria.
Pinus sylvestris—Scotch Fir.

Minoa murinata (euphorbiata).
Euphorbia Cyparissias—Cypress Spurge.
  „ *Paralias*—Sea Spurge.
  „ *Peplus*—Petty Spurge.

Scoria lineata (dealbata).
Polygonum aviculare—Knot-grass.
Also on various grasses.

Aplasta ononaria.
Ononis spinosa—Rest-harrow.

Sterrha sacraria.
Anthemis Cotula.
  „ *nobilis*—Chamomile.
  „ *arvensis*—Corn Chamomile.
Polygonum aviculare—Knot-grass. In confinement.
Rumex—Dock.

Lythria purpuraria.
 Polygonum aviculare—Knot-grass.

Aspilates strigillaria.
 Cytisus (Sarothamnus) scoparius—Broom.
 Calluna Erica—Ling ; Heather.

Aspilates ochrearia (citraria).
 Lotus corniculatus—Bird's-foot Trefoil.
  „ *pilosus*—Greater Bird's-foot Trefoil.
 Daucus Carota—Wild Carrot.
 Silene Otites—Spanish Catchfly.

Aspilates gilvaria.
 Achillea Millefolium—Yarrow ; Milfoil.
 Thymus Serpyllum—Thyme.
 Potentilla reptans—Trailing Tormentil.
 Medicago lupulina—Black Medick.
 Veronica serpyllifolia—Thyme-leaved Speedwell. In confinement.

Abraxas (Zerene) grossulariata.
 Prunus spinosa—Blackthorn.
 Ribes Grossularia—Gooseberry.
  „ *sanguinem*—Flowering Currant.
  „ *rubrum*—Red Currant.
 Cotyledon Umbilicus—Wall Pennywort.
 Sedum Telephium—Orpine.

Abraxas (Zerene) sylvata (ulmata).
 Ulmus montana—Wych Elm.
  „ *campestris*—Elm.

Ligdia adustata.
 Euonymus europæus—Spindle.

Lomaspilis marginata.
 Salix cinerea—(Sallow.)
  „ *Caprea*—(Sallow.)

Pachycnemia hippocastanaria.
 Erica Tetralix—Cross-leaved Heath.
  „ *cinerea*—Common Purple Heath.

Hybernia rupicapraria.
 Prunus spinosa—Blackthorn.
 Cratægus Oxyacantha—Whitethorn.
 Quercus Robur—Oak.

Hybernia leucophæaria.
 Acer Pseudo-platanus—Sycamore.
 Quercus Robur—Oak.

Hybernia aurantiaria.
Crataegus Oxyacantha—Whitethorn.
Betula alba—Birch.
Carpinus Betulus—Hornbeam.
Quercus Robur—Oak.

Hybernia marginaria (progemmaria).
Betula alba—Birch.
Carpinus Betulus—Hornbeam.
Quercus Robur—Oak.

Hybernia defoliaria.
Prunus spinosa—Blackthorn.
Crataegus Oxyacantha—Whitethorn.
Ulmus campestris—Elm.
Carpinus Betulus—Hornbeam.
Corylus Avellana—Hazel.
Quercus Robur—Oak.

Anisopteryx æscularia.
Tilia vulgaris—Lime.
Prunus spinosa—Blackthorn.
Crataegus Oxyacantha—Whitethorn.
Ulmus campestris—Elm.
Quercus Robur—Oak.

Cheimatobia brumata.
Prunus spinosa—Blackthorn.
Pyrus communis—Pear.
„ *Malus*—Apple.
Crataegus Oxyacantha—Whitethorn.
Betula alba—Birch.
Carpinus Betulus—Hornbeam.
Corylus Avellana—Hazel.
Quercus Robur—Oak.
Also on many other trees.

Cheimatobia boreata.
Betula alba—Birch.

Oporabia dilutata.
Prunus spinosa—Blackthorn.
Crataegus Oxyacantha—Whitethorn.
Carpinus Betulus—Hornbeam.
Quercus Robur—Oak.
Also on many other trees.

Oporabia filigrammaria.
Vaccinium Myrtillus—Whortleberry; Bilberry.
Calluna Erica—Ling; Heather.

Oporabia filigrammaria—*continued.*
   Erica Tetralix—Cross-leaved Heath.
     „ *cinerea*—Common Purple Heath.
   Salix Caprea—(Sallow.)
     „ *cinerea*—(Sallow.)

Oporabia autumnaria.
   Betula alba—Birch.

Larentia didymata.
   Anemone nemorosa—Wood Anemone.
   Chærophyllum temulum—Rough Chervil.
   Anthriscus sylvestris—Hare's Parsley.
   Vaccinium Myrtillus—Whortleberry; Bilberry.

Larentia multistrigaria.
   Galium verum—Lady's Bedstraw.
   Asperula odorata—Sweet Woodruff. In confinement.
   Also on other *Galia.*

Larentia cæsiata.
   Vaccinium Vitis-Idæa—Cow-berry.
     „   *Myrtillus*—Whortleberry; Bilberry. In confinement.
   Salix alba—White Willow. In confinement.
   Calluna Erica—Ling; Heather.
   Erica Tetralix—Cross-leaved Heath.
     „ *cinerea*—Common Purple Heath.

Larentia flavicinctata.
   Saxifraga granulata—White Meadow Saxifrage.
     „   *hypnoides*—Mossy Saxifrage.
     „   *aizoides*—Yellow Mountain Saxifrage.
   Calluna Erica—Ling; Heather.

Larentia salicata.
   Galium verum—Lady's Bedstraw.
   Asperula odorata—Sweet Woodruff. In confinement.
   Also on other *Galia.*

Larentia olivata.
   Galium Mollugo—Hedge Bedstraw.

Larentia viridaria (pectinitaria).
   Galium Mollugo—Hedge Bedstraw.
     „ *sexatile*—Heath Bedstraw.

Emmelesia affinitata (rivulata).
   Lychnis alba—Evening Campion. On the seeds.
     „ *diurna*—Red Campion. On the seeds.
     „ *Flos-cuculi*—Ragged Robin. On the seeds.

Emmelesia alchemillata.
  Galeopsis Ladanum—Red Hemp Nettle. On the seeds.
  „ *Tetrahit*—Hemp Nettle. On the seeds.

Emmelesia albulata.
  Rhinanthus Crista-galli—Yellow Rattle. On the seeds.

Emmelesia decolorata.
  Lychnis diurna—Red Campion. On the seeds.

Emmelesia tæniata.
  (Food-plant at present unknown).

Emmelesia unifasciata.
  Bartsia Odontites—Red Bartsia.

Emmelesia minorata (ericetata).
  (Food-plant at present unknown).

Emmelesia adæquata (blandiata).
  Euphrasia officinalis—Eye-bright.

Eupithecia venosata.
  Silene Cucubalus—White Campion. On the seeds.
  „ *maritima*—Sea Campion. On the seeds.
  „ *gallica*—English Catchfly. On the seeds.
  „ *acaulis*—Moss Campion. On the seeds.
  „ *nutans*—Nottingham Catchfly. On the seeds.
  Lychnis diurna—Red Campion. On the seeds.

Eupithecia consignata.
  Pyrus communis—Pear.
  „ *Malus*—Apple.

Eupithecia linariata.
  Linaria vulgaris—Yellow Toad-flax. On the flowers and seeds.

Eupithecia oblongata (centaureata).
  Pimpinella Saxifraga—Common Burnet Saxifrage. On the flowers.
  „ *major*—Great Burnet Saxifrage. On the flowers.
  Silaus pratensis—Meadow Saxifrage. On the flowers.
  Scabiosa Columbaria—Small Scabious. On the flowers.
  Eupatorium cannabinum—Hemp Agrimony. On the flowers.
  Solidago Virgaurea—Golden-rod.
  Achillea Millefolium—Yarrow; Milfoil.
  Senecio vulgaris—Groundsel. On the flowers.
  „ *erucifolius*—Hoary Ragwort. On the flowers.
  „ *Jacobæa*—Ragwort. On the flowers.
  Centaurea nigra—Knap-weed. On the flowers.
  Campanula glomerata—Clustered Bell-flower. On the flowers.

Eupithecia pulchellata.
  Digitalis purpurea—Foxglove. On the flowers.
Eupithecia succentauriata.
  Achillea millefolium—Yarrow ; Milfoil.
  Artemisia Absinthium—Common Wormwood. On the flowers.
  " *vulgaris*—Mugwort.
  Senecio Jacobæa—Ragwort. On the flowers.
  Anthemis nobilis—Chamomile.
Eupithecia subfulvata.
  Achillea Millefolium—Yarrow ; Milfoil.
Eupithecia scabiosata (subumbrata).
  Daucus Carota—Wild Carrot.
  Galium Mollugo—Hedge Bedstraw.
  Scabiosa arvensis—Field Scabious. On the flowers.
  Centaurea nigra—Knap-weed. On the flowers.
  Leontodon hispidus—Rough Hawk-bit. On the flowers.
  Gentiana Amazella—Autumnal Gentian. On the flowers.
  " *campestris*—Field Gentian. On the flowers.
  Origanum vulgare—Marjoram. On the flowers.
  Prunella vulgaris—Self-heal. On the flowers.
Eupithecia pernotata.
  Solidago Virgaurea—Golden-rod.
Eupithecia plumbeolata.
  Melampyrum pratense—Cow-wheat. On the flowers.
Eupithecia isogrammaria (haworthiata).
  Clematis Vitalba—Traveller's Joy. In the flower-buds.
Eupithecia pygmæata (palustraria).
  Cerastium arvense—Field Mouse-ear Chickweed.
  Stellaria Holostea—Greater Stitchwort. On the flowers.
Eupithecia helveticaria.
  Juniperus communis—Juniper.
Eupithecia helveticaria v. (?) arceuthata.
  Juniperus communis—Juniper.
Eupithecia egenaria.
  (Food-plant unknown ; abroad on *Tunica saxifraga*).
Eupithecia satyrata.
  Hypericum perforatum—Perforated St. John's Wort. On the flowers.
  Galium Mollugo—Hedge Bedstraw.
  Scabiosata succisa—Devil's-bit Scabious. On the flowers.
  " *arvensis*—Field Scabious. On the flowers.
  Centaurea nigra—Knap-weed. On the flowers.
  Leontodon hispidus—Rough Hawk-bit. On the flowers.

Eupithecia satyrata—*continued.*
Gentiana Amarella—Autumnal Gentian. On the flowers.
    ,,   *campestris*—Field Gentian. On the flowers.
Origanum vulgare—Marjoram. On the flowers.
Prunella vulgaris—Self-heal. On the flowers.

Eupithecia castigata.
Silene cucubalus—White Campion. On the seeds. Polyphagous.
Senecio Jacobæa—Ragwort (?).
Feeds on nearly every tree, shrub and plant.

Eupithecia jassioneata.
Jasione montana—Sheep's Scabious.

Eupithecia trisignaria.
Angelica sylvestris—Wild Angelica. On the flowers.

Eupithecia virgaureata.
Pimpinella Saxifraga—Common Burnet Saxifrage. On the flowers.
    ,,   *major*—Great Burnet Saxifrage. On the flowers.
Solidago Virgaurea—Golden-rod.
Senecio Jacobæa—Ragwort. On the flowers.
    ,,   *palustris*—Marsh Flea-wort. In confinement.

Eupithecia fraxinata.
Fraxinus excelsior—Ash.

Eupithecia extensaria.
Artemisia Absinthium—Common Wormwood. On the flowers.

Eupithecia pimpinellata.
Pimpinella Saxifraga—Common Burnet Saxifrage. On the flowers.

Eupithecia valerianata (viminata).
Valeriana officinalis—Great Wild Valerian. On the flowers.

Eupithecia pusillata.
Juniperus communis—Juniper.

Eupithecia irriguata.
Quercus Robur—Oak.
Fagus sylvatica—Beech.

Eupithecia campanulata (denotata).
Campanula Trachelium—Canterbury Bells. On the seeds. Also on most *Campanulæ*.

Eupithecia innotata.
Artemisia Absinthium—Common Wormwood. On the flowers.

Eupithecia indigata.
Juniperus communis—Juniper.
Pinus—Fir.
Also on Cypress.

Eupithecia constrictata.
Thymus Serpyllum—Thyme. On the flowers.

Eupithecia nanata
Calluna Erica—Ling; Heather.
Erica Tetralix—Cross-leaved Heath.
„ *cinerea*—Common Purple Heath.

Eupithecia subnotata.
Chenopodium Vulvaria—Stinking Goose-foot; Orach. On the flowers and seeds.
„ *album*—White Goose-foot. On the flowers and seeds.
„ *Bonus-Henricus*—Good King Henry. On the flowers and seeds.
Beta maritima—Beet.
Atriplex patula—Orach.
„ *laciniata*—Sea Orach.

Eupithecia vulgata.
Cratægus Oxyacantha—Whitethorn.
Solidago Virgaurea—Golden-rod.
Senecio Jacobæa—Ragwort (?).
Salix alba—White Willow.

Eupithecia albipunctata.
Angelica sylvestris—Wild Angelica. On the flowers.
Heracleum Sphondylium—Cow-parsnep. On the flowers.

Eupithecia expallidata.
Solidago Virgaurea—Golden-rod.
Senecio Jacobæa—Ragwort. On the flowers.

Eupithecia absinthiata.
Eupatorium cannabinum—Hemp Agrimony. On the flowers.
Solidago Virgaurea—Golden-rod.
Achillea Millefolium—Yarrow; Milfoil.
Tanacetum vulgare—Tansy.
Artemisia Absinthium—Common Wormwood. On the flowers.
„ *vulgaris*—Mugwort. On the flowers.
Senecio vulgaris—Groundsel. On the flowers.
„ *erucifolius*—Hoary Ragwort. On the flowers.
„ *Jacobæa*—Ragwort. On the flowers.

Eupithecia minutata.
Scabiosa succisa—Devil's-bit Scabious. On the flowers.
Calluna Erica—Ling ; Heather.
Erica Tetralix—Cross-leaved Heath.
„ *cinerea*—Common Purple Heath.

Eupithecia assimilata.
Ribes nigrum—Black Currant.
Humulus Lupulus—Hop.

Eupithecia tenuiata.
Salix cinerea—(Sallow).
„ *Caprea*—(Sallow).

Eupithecia subciliata.
Acer campestre—Maple. On the flowers.

Eupithecia lariciata.
Pinus abies—Spruce Fir.
Larix Europœa—Larch.

Eupithecia abbreviata.
Quercus Robur—Oak.

Eupithecia dodoneata.
Cratægus Oxyacantha—Whitethorn.
Quercus Robur—Oak.

Eupithecia exiguata.
Berberis vulgaris—Barbary.
Cratægus Oxyacantha—Whitethorn.
Ribes nigrum—Black Currant.
Fraxinus excelsior—Ash.
Alnus glutinosa—Alder.
Salix cinerea—(Sallow.)
,, *Caprea*—(Sallow.)

Eupithecia ultimaria.
Tamarix gallica—Tamarisk.

Eupithecia sobrinata.
Juniperus communis—Juniper.

Eupithecia togata.
Pinus abies—Spruce Fir. In the cones.

Eupithecia pumilata.
Clematis Vitalba—Traveller's Joy. On the flowers.
Anthriscus sylvestris—Hare's Parsley. On the flowers.
Convolvulus arvensis—Field Bindweed.

Eupithecia coronata.
Clematis Vitalba—Traveller's Joy. On the flowers.
Angelica sylvestris—Wild Angelica. On the flowers.
Eupatorium cannabinum—Hemp Agrimony. On the flowers.
Solidago Virgaurea—Golden-rod.

Eupithecia rectangulata.
Pyrus communis—Pear. In the buds.
„ *Malus*—Apple. In the buds.

Eupithecia debiliata.
Vaccinium Myrtillus—Whortleberry; Bilberry.

Collix sparsata.
Lysimachia vulgaris—Loosestrife.

Lobophora sexalisata (sexalata).
Salix alba—White Willow.
„ *Caprea*—(Sallow).

Lobophora halterata (hexapterata).
Salix cinerea—(Sallow).
„ *Caprea*—(Sallow).
Populus tremula—Aspen.

Lobophora viretata.
Pyrus Aucuparia—Mountain Ash.
Ligustrum vulgare—Privet.
Ilex Aquifolium—Holly.
Viburnum Opulus—Guelder Rose.
Rhamnus frangula—Alder Buckthorn.
Acer Pseudo-platanus—Sycamore.
Hedera Helix—Ivy.

Lobophora carpinata (lobulata).
Lonicera Periclymenum—Honeysuckle.
Salix Caprea—(Sallow.)

Lobophora polycommata.
Lonicera Periclymenum—Honeysuckle.
Fraxinus excelsior—Ash.

Thera juniperata.
Juniperus communis—Juniper.

Thera simulata.
Juniperus communis—Juniper.

Thera variata.
Pinus sylvestris—Scotch Fir.

Thera firmata.
Pinus—Fir.

Hypsipetes ruberata.
Salix alba—White Willow.
„ *cinerea*—(Sallow).
„ *Caprea*—(Sallow).

Hypsipetes trifasciata (impluviata).
Alnus glutinosa—Alder.

Hypsipetes sordidata (elutata).
Vaccinium Myrtillus—Whortleberry; Bilberry.
Calluna Erica—Ling; Heather.
Alnus glutinosa—Alder.
Salix cinerea—(Sallow).
„ *Caprea*—(Sallow).

Melanthia bicolorata (rubiginata).
Prunus spinosa—Blackthorn.
Alnus glutinosa—Alder.

Melanthia ocellata.
Galium verum—Lady's Bedstraw.

Melanthia albicillata.
Rubus fruticosus—Bramble.
„ *Idæus*—Raspberry.
Fragaria vesca—Wild Strawberry.

Melanippe hastata.
Myrica Gale—Sweet Gale.
Betula alba—Birch.

Melanippe tristata.
Galium Mollugo—Hedge Bedstraw.

Melanippe procellata.
Clematis Vitalba—Traveller's Joy.

Melanippe unangulata.
Stellaria media—Chickweed.

Melanippe rivata.
Galium Mollugo—Hedge Bedstraw.

Melanippe sociata (subtristata).
Galium Mollugo—Hedge Bedstraw.

Melanippe montanata.
Primula acaulis—Primrose.

Melanippe galiata.
Galium verum—Lady's Bedstraw.
„ *Mollugo*—Hedge Bedstraw.

Melanippe fluctuata.
Brassica Napus—Rape.
Tropæolum majus.
Also on varieties of cabbage, and other garden plants.

Anticlea cucullata (sinuata).
Galium verum—Lady's Bedstraw.

Anticlea rubidata.
Galium verum—Lady's Bedstraw.
„ *Mollugo*—Hedge Bedstraw.

Anticlea badiata.
Rosa spinosissima—Burnet-rose.
„ *canina*—Dog Rose.

Anticlea nigrofasciaria (derivata).
Rosa canina—Dog Rose.
Lonicera Periclymenum—Honeysuckle.

Anticlea berberata.
Berberis vulgaris—Barberry.

Coremia munitata.
Senecio vulgaris—Groundsel.

Coremia designata (propugnata).
Brassica Napus—Rape.
Also on cabbage.

Coremia ferrugata.
Stellaria media—Chickweed.
Nepeta Glechoma—Ground Ivy.

Coremia unidentaria.
Galium verum—Lady's Bedstraw.
Asperula odorata—Sweet Woodruff.

Coremia quadrifasciaria.
Cratægus Oxyacantha—Whitethorn.
Lamium album—White Dead-nettle.
Also on various low plants.

Camptogramma bilineata.
Poa annua—Annual Meadow-grass.
Also on other grasses.

Camptogramma fluviata.
Senecio vulgaris—Groundsel.
Polygonum Persicaria—Common Persicaria.

Phibalapteryx tersata.
Clematis Vitalba—Traveller's Joy.

Phibalapteryx lapidata.
(Evergreen oak—abroad.)

Phibalapteryx vittata (lignata).
Galium verum—Lady's Bedstraw.
„ *Mollugo*—Hedge Bedstraw.
„ *palustre*—Water Bedstraw.
„ *saxatile*—Heath Bedstraw. In confinement.

Phibalapteryx polygrammata.
Galium verum—Lady's Bedstraw.
Also on other *Galia*.

Phibalapteryx vitalbata.
Clematis Vitalba—Traveller's Joy.

Triphosa dubitata.
Rhamnus catharticus—Buckthorn.

Eucosmia certata.
Berberis vulgaris—Barberry.

Eucosmia undulata.
Vaccinium Myrtillus—Whortleberry; Bilberry.
Salix cinerea—(Sallow).
„ *Caprea* (Sallow).

Scotosia vetulata.
Rhamnus catharticus—Buckthorn.

Scotosia rhamnata.
Rhamnus catharticus—Buckthorn.

Cidaria siterata (psittacata).
Tilia vulgaris—Lime.
Rosa canina—Dog Rose.
Pyrus Malus—Apple.
Fraxinus excelsior—Ash.
Quercus Robur—Oak.

Cidaria miata.
Betula alba—Birch.
Alnus glutinosa—Alder.
Quercus Robur—Oak.

Cidaria picata.
Stellaria media—Chickweed. In confinement.
Galium Mollugo—Hedge Bedstraw.

Cidaria corylata.
Tilia vulgaris—Lime.
Prunus spinosa—Blackthorn.

Cidaria sagittata.
Thalictrum flavum—Yellow Meadow Rue.
In confinement on *Thalictrum glaucum* and *T. aquilegifolium.*

Cidaria truncata (russata).
Fragaria vesca—Wild Strawberry.
Cratægus Oxyacantha—Whitethorn.
Betula alba—Birch.
Salix cinerea—(Sallow).
„ *Caprea*—(Sallow).

Cidaria immanata.
Fragaria vesca—Wild Strawberry.
Vaccinium Myrtillus—Whortleberry; Bilberry.

Cidaria suffumata.
  Galium Mollugo—Hedge Bedstraw.
Cidaria reticulata.
  Impatiens Noli-tangere.—Do-not-touch-me.
Cidaria silaceata.
  Epilobium hirsutum—Great Hairy Willow-herb.
  Circæa lutetiana—Enchanter's Nightshade.
  Populus tremula—Aspen.
  Also on other *Epilobia.*
Cidaria prunata.
  Ribes Grossularia—Gooseberry.
  ,,   *rubrum*—Red Currant.
  ,,   *nigrum*—Black Currant.
Cidaria testata.
  Betula alba—Birch.
  Salix cinerea—(Sallow).
  ,,   *Caprea*—(Sallow).
  Populus tremula—Aspen.
Cidaria populata.
  Vaccinium Myrtillus—Whortleberry; Bilberry.
  ,,   *Vitis-Idæa*—Cow-berry.
  Salix cinerea—(Sallow).
  ,,   *Caprea*—(Sallow).
  ,,   *alba*—White Willow. In confinement.
Cidaria fulvata.
  Rosa spinosissima—Burnet Rose.
  ,,   *canina*—Dog Rose.
Cidaria dotata (pyraliata).
  Cratægus Oxyacantha—Whitethorn.
  Galium Mollugo—Hedge Bedstraw.
Cidaria associata (dotata).
  Ribes rubrum—Red Currant.
  ,,   *nigrum*—Black Currant.
Pelurga comitata.
  Chenopodium Vulvaria—Stinking Goose-foot; Orach.
  ,,   *album*—White Goose-foot.
  ,,   *Bonus-Henricus*—Good King Henry.
Eubolia cervinata.
  Malva sylvestris—Mallow.
Eubolia limitata (mensuraria).
  Poa pratensis.
  Also on other grasses.

Eubolia plumbaria (palumbaria).
 Genista anglica—Needle Furze.
 Cytisus (Sarothamnus) scoparius—Broom.
 Lotus corniculatus—Bird's-foot Trefoil.
 Erica Tetralix—Cross-leaved Heath.
 „ cinerea—Common Purple Heath.

Eubolia bipunctaria.
 Lotus corniculatus—Bird's-foot Trefoil.

* **Eubolia mœniata.**
 Cytisus (Sarothamnus) scoparius—Broom.

Mesotype virgata (lineolata).
 Galium verum—Lady's Bedstraw.
 „ saxatile—Heath Bedstraw. In confinement.

Carsia paludata (imbutata).
 Vaccinium Oxycoccos—Cranberry.
 „ Vitis-Idæa—Cow-berry.

Anaitis plagiata.
 Hypericum perforatum—Perforated St. John's Wort.

Lithostege griseata (nivearia).
 Sisymbrium officinale—Common Hedge Mustard.
 Erisymum cheiranthoides—Worm-seed Treacle Mustard.

Chesias spartiata.
 Cytisus (Sarothamnus) scoparius—Broom.

Chesias rufata (obliquaria).
 Cytisus (Sarothamnus) scoparius—Broom.

Tanagra atrata (chærophyllata).
 Conopodium (Bunium) denudatum—Common Pig-nut. On the blossoms.
 Chærophyllum temulum—Rough Chervil.

FOOD-PLANTS AND LARVÆ.

Clematis Vitalba: TRAVELLER'S JOY.—47.
   Lithosia lurideola (complanula).—Buckl.
   Angerona prunaria.
   Geometra vernaria.—Newm., Sta.
   Acidalia strigilaria (strigilata).—Ent. xiv. 215.
     „   *immutata.*—In confinement, after hybernation.
   Eupithecia isogrammaria (haworthiata).—In the buds.
     „   *pumilata.*—On the flowers. Newm., Sta.
     „   *coronata.*—On the flowers. Newm., Sta.
   Melanippe procellata.—Newm.
   Phibalapteryx tersata.—Newm., Sta.
     „   *vitalbata.*—Newm., Sta.

Thalictrum flavum: YELLOW MEADOW RUE.—65.
   Cidaria sagittata.—Newm., Ent. Mag. iii. 110; xii. 113.

Anemone nemorosa: WOOD ANEMONE.—105.
   Larentia didymata.

Ranunculus acris: MEADOW CROWFOOT.—111.
   Tryphæna orbona (subsequa).—Ent. Mag. ix. 57.

Ranunculus repens: CREEPING CROWFOOT.—111.
   Tryphæna orbona (subsequa).—Ent. Mag. ix. 57.

Ranunculus bulbosus: BULBOUS BUTTERCUP.—97.
   Anchocelis pistacina.—Also on other *Ranunculi.* Newm.
   Trigonophora flammea (empyrea).—Ent. xvi.i. 162.

Ranunculus Ficaria: PILE-WORT.—105.
   Trigonophora flammea (empyrea).—Newm. Sta.

Helleborus fœtidus:—STINKING HELLEBORE.—13.
   Sesia ichneumoniformis.—In the stems. Newm.

Berberis vulgaris: BARBARY.—77.
   Eupithecia exiguata.—Newm.
   Anticlea berberata.—Newm., Sta.
   Eucosmia certata.—Newm., Sta.

Papaver Rhœas: RED POPPY.—102.
   Agrotis aquilina.—Newm.

Nasturtium officinale : WATER-CRESS.—109.
  Pieris rapæ.—Buckl., Ent., xxii. 18.
  „ *napi.*—Newm., Buckl.
  Euchloë cardamines.—Buckl., Ent. xxii. 18.

Barbarea vulgaris : WINTER CRESS.—95.
  Pieris napi.—Buckl.
  Euchloë cardamines.—On the seeds. Newm.

Barbarea præcox : EARLY WINTER CRESS.
  Pieris napi.—Ent. xiii. 207.

Arabis perfoliata : ROCK CRESS.—38.
  Euchloë cardamines.—Newm., Buckl.

Cardamine pratensis : CUCKOO-FLOWER.—111.
  Pieris napi.—Ent. Mag. xxiii. 20., Buckl.
  Euchloë cardamines.—Ent. Mag. xxiii. 20., Buckl.

Cardamine impatiens : NARROW-LEAVED BITTER-CRESS.—26.
  Euchloë cardamines.—Sta.

Hesperis matronalis : DAME'S VIOLET.
  Euchloë cardamines.—Buckl., Newm.

Sisymbrium officinale : COMMON HEDGE MUSTARD.—107.
  Euchloë cardamines.—Buckl.
  Pieris rapæ.—Buckl.

Sisymbrium Sophia : FLIXWEED.—64.
  Lithostege griseata (nivearia).—Newm.

Sisymbrium Alliaria : JACK-BY-THE-HEDGE.—94.
  Pieris napi.—Newm.
  Euchloë cardamines.—Buckl., Newm.

Erysimum cheiranthoides: WORM-SEED TREACLE-MUSTARD.—37.
  Lithostege griseata (nivearia).—Newm.

Brassica Napus : RAPE.
  Pieris brassicæ.
  „ *rapæ.*
  „ *napi.*—Sta.
  Agrotis exclamationis.—Newm., Sta.
  Melanippe fluctuata.—Newm., Sta.
  Coremia designata (propugnata).—Newm., Sta.

Brassica Sinapis : CHERLOCK ; WILD MUSTARD.—111.
  Pieris rapæ.—Buckl.
  Euchloë cardamines.—Ent. Mag. xxiii. 20., Buckl.
  Agrotis segetum.—Newm.

FOOD-PLANTS AND LARVÆ.

Cakile maritima : SEA ROCKET.—59.
  Pieris napi.—Buckl.

Reseda Luteola : DYER'S ROCKET.—94.
  Pieris daplidice.—Buckl., Newm., Sta.

Reseda lutea : WILD MIGNONETTE.—53.
  Pieris daplidice.—Newm., Sta.
  „   *rapæ.*—Ent. Mag. xxiii. 20.
  Heliothis armigera.—Sta.

Helianthemum Chamæcistus : ROCK ROSE.—88.
  Lycæna astrarche (agestis).—Buckl.
    „   „   *v. artaxerxes.*—Buckl., Newm., Ent. Mag. v. 176.
  Ino geryon.—Newm., Buckl.
  Agrotis ashworthii.—Newm., Ent. xxiii. 6.
  Gnophos obscuraria (obscurata).—Newm., Sta.
  Acidalia holosericata.—Ent. Mag. v. 96.   In confinement.

Viola odorata : SWEET VIOLET.—79.
  Argynnis paphia.—Newm.
    „   *latona (lathonia).*—Newm.
  Nemeophila plantaginis.—Newm.
  Rusina tenebrosa.—Newm.
  Polia xanthomista.—Newm.

Viola canina : DOG VIOLET.—77.
  Argynnis paphia.—Newm., Sta.
    „   *aglaia.*—Newm., Sta.
    „   *adippe.*—Newm., Sta.
    „   *latona (lathonia).*—Newm.
    „   *selene.*—Newm., Sta., Ent. Mag. vii. 115.
    „   *euphrosyne.*—Newm., Sta., Ent. Mag. v. 126.
  *   „   *dia.*
  Nemeophila plantaginis.—Newm.
  Rusina tenebrosa.—Newm.
  Polia xanthomista.—Newm.

Viola tricolor : HEARTSEASE.—110.
  Argynnis adippe.—Sta.
  *   „   *niobe.*—Newm.
    „   *latona (lathonia).*—Newm., Sta.
  *   „   *dia.*
  Nemeophila plantaginis.—Newm.
  Rusina tenebrosa.—Newm.
  Polia xanthomista.—Newm.

Viola Curtisii Forster.—17.
  Agrotis cursoria.—Ent. Mag. ix. 14.

Polygala vulgaris.—COMMON MILK-WORT.—59.
  Phytometra viridaria (ænea).—Newm., Ent. Mag. x. 139.
Dianthus Armeria: DEPTFORD PINK.—45.
  Acidalia marginepunctata (promutata). Also on other *Dianthi.* Sta.
Silene Cucubalus: WHITE CAMPION.—102.
  Neuria reticulata (saponariæ).—Newm., Sta.
  Dianthœcia nana (conspersa).—Sta., Ent xvi. 248. On the seeds.
  „   *albimacula.*—Ent. Mag. xi. 16. On the seeds; in confinement.
  „   *cæsia.*—Newm. On the seeds.
  „   *irregularis.* In confinement.
  „   *capsincola.*—Ent. xvi. 248. On the seeds.
  „   *cucubali.*—Newm., Sta. On the seeds.
  „   *carpophaga.*—Newm., Sta. On the seeds.
  „   *capsophila.*—Newm. On the seeds; in confinement.
  Calocampa exoleta.—Newm.
  Polia xanthomista.—Newm.
  Eupithecia venosata.—Newm., Sta. On the seeds.
  „   *castigata.* On the flowers.
Silene maritima: SEA CAMPION.—75.
  Aporophyla australis.—Ent. xix. 283.
  Neuria reticulata (saponariæ).—Newm., Sta
  Dianthœcia nana (conspersa).—Ent. xvi. 248. On the seeds.
  „   *albimacula.*—Ent. Mag. xi. 16. In confinement; on the seeds.
  „   *capsincola.*—Ent. xvi. 248. On the seeds.
  „   *cucubali.*—Ent. xvi. 247. On the seeds.
  „   *capsophila.*—Newm.,Ent. Mag. xii. 138. On the seeds.
  „   *cæsia.*— Ent. Mag. xii. 138. On the seeds.
  Polia xanthomista.—Newm., Ent. Mag. x. 89.
  Acidalia perochraria (ochrearia).
  „   *emutaria.* In confinement.
  Eupithecia venosata.—Ent. xviii. 111., Sta. On the seeds.
Silene gallica: ENGLISH CATCHFLY.—57.
  Neuria reticulata (saponariæ).—Newm., Sta.
  Polia xanthomista.—Newm.
  Eupithecia venosata.—Ent. xviii. 111., Sta. On the seeds.
Silene acaulis: MOSS CAMPION.—21.
  Zygæna exculans.—Buckl., Ent. Mag. xx. 151.
  Noctua festiva v. conflua.—Newm.
  Polia xanthomista. Newm.
  Eupithecia venosata.—Ent. xviii. 111., Sta. On the seeds.

Silene Otites: SPANISH CATCHFLY.—3.
    Dianthœcia irregularis.—Ent. Mag. xvii. 125, Ent. ix. 232.
    Heliothis dipsacea.—Ent. Mag. xi. 257, xvii. 125. On the
        flowers and seeds.
    Aspilates ochrearia (citraria).

Silene nutans: NOTTINGHAM CATCHFLY.—15.
    Dianthœcia albimacula.—Newm., Sta.
      ,,   *conspersa.*—Sta.
    Polia xanthomista.— Newm.
    Eupithecia venosata.—Ent. xviii. 111., Sta. On the seeds.

Lychnis alba: EVENING CAMPION.—94.
    Macroglossa fuciformis.—Newm.
    Dianthœcia nana (conspersa).—Ent. xvi. 248. On the seeds.
      ,,   *capsincola.*—Ent. xvi. 248, Newm. On the seeds.
      ,,   *cucubali.*—Ent. xvi. 248. On the seeds.
      ,,   *carpophaga.*—Ent. xvi. 248. On the seeds.
    Heliothis dipsacea.
    Emmelesia affinitata.—Sta. On the seeds.

Lychnis diurna: RED CAMPION.—111.
    Macroglossa fuciformis.—Newm.
    Dianthœcia nana (conspersa).—Ent. xvi. 248. On the seeds.
      ,,   *albimacula.*—Ent. Mag. xi. 16. In confinement.
            On the seeds.
      ,,   *capsincola.*—Ent. xvi. 248, Sta. On the seeds.
      ,,   *cucubali.*—Ent. xvi. 248. On the seeds.
      ,,   *carpophaga.*—Ent. xvi. 248. On the seeds.
    Emmelesia decolorata.—Sta. On the flowers.
      ,,   *affinitata.*— Sta. On the seeds.
    Eupithecia venosata.—Newm. On the seeds.

Lychnis Flos-cuculi: RAGGED ROBIN.—111.
    Macroglossa fuciformis.—Newm.
    Dianthœcia nana (conspersa).—Newm., Sta. On the seeds.
      ,,   *cucubali.* On the seeds.
    Emmelesia affinitata. Sta. On the seeds.

Lychnis Githago: CORN-COCKLE.—97.
    Tryphœna pronuba.—Ent. Mag. xvii. 135.

Cerastium glomeratum: MOUSE-EAR CHICKWEED.—110.
    Heliaca (Heliodes) tenebrata (arbuti). Ent. Mag. xix. 39.

Cerastium arvense: FIELD MOUSE-EAR CHICKWEED.—69.
    Heliaca (Heliodes) tenebrata (arbuti).—Newm., Sta.
    Eupithecia pygmæata.—Ent. xviii. 145.

Stellaria Holostea: GREATER STITCHWORT.—106.
    Eupithecia pygmæata.—Ent. xviii. 145, Ent. Mag. ix. 42 ; ix.
        65. On the flowers.

Stellaria media: CHICKWEED.—112.
   Arctia villica.—Newm.
   Spilosoma mendica.—Newm.
   Leucania lithargyria.—Sta.
      „   *albipuncta.*
   Xylophasia hepatica.—Newm.
   Aporophyla australis.—Ent. Mag. vi. 13.
   Caradrina alsines.—Newm., Sta.
      „   *taraxaci (blanda).*—Newm.
      „   *quadripunctata (cubicularis).*—Sta.
   Agrotis aquilina.—Newm.
      „   *præcox.*—Newm.
   Tryphæna comes (orbona).—Newm.
   Tryphæna pronuba.
   Noctua stigmatica (rhomboidea).—Newm.
   Mania maura.—Sta.
   Polia flavicincta (flavocincta).—Newm., Sta.
   Epunda nigra.—Sta.
   Hadena genistæ.—Newm. In confinement.
   Melanippe unangulata.—Newm.
   Coremia ferrugata.—Sta.
   Cidaria picata.—Newm. In confinement.
   Acidalia contiguaria.—Ent. xi. 242.
      „   *trigeminata.* In confinement.
      „   *immutata.* In confinement.
      „   *emutaria.* In confinement.
      „   *degenaria.* In confinement.

Arenaria peploides: SEA SANDWORT. —68.
   Agrotis cursoria.—Ent. Mag. ix. 14.

Arenaria (Cherleria) sedoides: MOSSY CYPHEL.—8.
   Zygæna exulans.—Buckl.

Arenaria: SANDWORT.
   Heliothis peltigera.—Newm., Sta.

Lepigonum (Spergularia) rubrum: PURPLE SANDWORT. 92.
   Heliothis peltigera.—Newm., Sta.

Tamarix gallica: TAMARISK.
   Eupithecia ultimaria.

Hypericum perforatum: PERFORATED ST. JOHN'S WORT.— 97.
   Cloantha polyodon (perspicillaris). Newm., Sta.
   Also on other *Hyperica.*
   Eupithecia satyrata.—Sta. On the flowers.
   Anaitis plagiata —Newm., Sta. Also on other *Hyperica.*

Malva sylvestris: MALLOW—96.
  Vanessa cardui.—Buckl.
  Tryphæna interjecta.—Newm.
  Eubolia cervinata.—Newm., Sta.

Tilia vulgaris: LIME.
  Smerinthus tiliæ.—Newm., Sta., Buckl.
  Arctia caia (caja).—Ent. Mag. xxiii. 262.
  *Laria L-nigrum.—Sta.
  Dasychira pudibunda.—Newm.
  Orgyia antiqua.
  Pœcilocampa populi.—Sta.
  Drepana harpagula (sicula).—Sta.
  Phalera bucephala.—Newm., Sta.
  Acronycta psi.
  „   *alni.*—Sta.
  Xanthia citrago—Newm., Sta., Ent. xii. 163.
  Xylina socia (petrificata).—Newm., Sta.
  Asteroscopus sphinx (cassinea).—Sta.
  Biston hirtaria.—Newm.
  Amphidasys betularia.—Newm.
  Anisopteryx æscularia.—Newm.
  Cidaria siterata (psittacata).—Sta.
  „   *corylata.*—Sta.

Tilia cordata (parvifolia).
  Drepana harpagula (sicula).—Buckl.

Erodium cicutarium: HEMLOCK STORK'S-BILL.—99.
  Lycæna astrarche (medon).—Newm., Sta.

Impatiens Noli-tangere: DO-NOT-TOUCH-ME.—24.
  Sphinx convolvuli.—Sta.
  Hadena rectilinea.—Ent. Mag. xxii. 91.
  Cidaria reticulata.

Tropæolum majus: (NOT BRITISH).
  Pieris brassicæ.—Buckl.
  „   *rapæ.*—Buckl.
  Melanippe fluctuata.—Newm.

Tropæolum canariense: (NOT BRITISH.)
  Pieris rapæ.—Buckl.

Ilex Aquifolium: HOLLY.—101.
  Lycæna argiolus.—Buckl., Newm., Sta.   On the flowers and berries.
  Sphinx ligustri.—Buckl., (Albin), Ent. ix. 275.
  Lobophora viretata.—Ent. x. 98; xix. 255; xx. 261.

Euonymus europæus: SPINDLE.—71.
  Boarmia repandata.—Sta.
  Ligdia adustata.—Newm., Sta.

Rhamnus catharticus : BUCKTHORN.—55.
   Gonopteryx rhamni.—Newm., Sta., Buckl.
   Lycæna argiolus.
   Lithosia luridcola (complanula).—Buckl. "Said to feed on lichen."
   Lasiocampa quercifolia.—Ent. xiii. 185.
   Triphosa dubitata.—Newm., Sta.
   Scotosia vetulata.—Newm.
      „ *rhamnata.*—Newm.

Rhamnus Frangula : ALDER BUCKTHORN.—58.
   Gonopteryx rhamni.—Newm., Sta., Buckl.
   Lycæna argiolus.—Buckl., Sta.
   Lithosia lurideola (complanula).—Buckl.
   Lobophora viretata.—Ent. x. 98.

Acer Pseudo-platanus : SYCAMORE.
   Zeuzera pyrina (æsculi). In the wood.
   Lophopteryx cuculla (cucullina).—Buckl.
   Ptilophora plumigera.—Ent. Mag. ix. 248.
   Notodonta dictæoides.—Buckl.
   Acronycta aceris.—Newm., Sta.
   Eugonia (Ennomos) autumnaria (aeniaria).
   Amphidasys betularia.—Ent. xix. 254.
   Zonosoma (Ephyra) annulata (omicronaria).—Ent. x. 137.
   Hybernia leucophæaria.
   Lobophora viretata—Ent. x. 98.

Acer campestre : MAPLE.—62.
   Ptilophora plumigera.—Newm., Sta., Buckl., Ent. Mag. vii. 210.
   Lophopteryx camelina.—Newm.
      „ *cuculla (cucullina)*—Newm., Sta., Buckl.
   Zonosoma (Ephyra) annulata (omicronaria).—Newm., Sta., Ent. x. 137.
   Asthena luteata.—Ent. Mag. xxiii. 109.
   Eupithecia subciliata.—Newm., Ent. Mag. ix. 16. On the flowers.

Æsculus Hippocastanum : HORSE CHESTNUT.
   Zeuzera pyrina (æsculi).—Sta.
   Acronycta aceris.—Newm., Sta.

Vitis vinifera : VINE.
   Deilephila livornica.—Newm., Sta.
   Chærocampa celerio.—Buckl., Newm., Sta.
      „ *elpenor.*—Buckl., Sta. In confinement.
*   „ *nerii.*
      „ *porcellus.*—Ent. xii. 250.

Genista anglica: NEEDLE FURZE.—86.
 Pseudoterpna pruinata (cytisaria).—Newm, Sta., Ent. x. 74.
 Eubolia plumbaria (palumbaria).—Newm.

Genista tinctoria: DYER'S GREEN-WEED.—73.
 Thecla rubi—Buckl.
 Dasydia obfuscaria (obfuscata).—Newm., Sta.

Cytisus (Sarothamnus) scoparius: BROOM.—109.
 Thecla rubi.—Buckl., Ent. Mag. vii. 232.
 Bombyx trifolii.—Sta.
 Noctua glareosa.—Newm.
 Hadena pisi.—Newm., Sta.
  „ *thalassina.*—Sta.
 Angerona prunaria.—Newm.
 Pseudoterpna pruinata (cytisaria).—Newm., Sta.
 Acidalia imitaria.—Newm.
 Fidonia limbaria (conspicuata).—Newm., Sta.
 Aspilates strigillaria.—Sta.
 Eubolia plumbaria (palumbaria).—Newm.
 * „ *mœniata.*—Newm., Sta.
 Chesias spartiata.—Newm., Sta., Ent. Mag. vii. 261.
  „ *rufata (obliquaria).*—Newm., Sta., Ent. Mag. vii. 261.

Ononis spinosa: REST-HARROW.—69.
 Lycœna icarus (alexis).—Newm., Buckl., Ent. Mag. iii. 91.
 Calocampa exoleta.—Newm.
 Heliothis umbra (marginatus).—Newm., Sta.
  „ *peltigera.*—Newm., Sta.
  „ *dipsacea.*—Ent. Mag. xi. 257. On the flowers and seeds.
 Aplasta ononaria.—Newm.
 Acidalia ochrata.

Ulex europæus: FURZE; GORSE.—110.
 Pseudoterpna pruinata (cytisaria).—Ent. x. 74.

Medicago sativa: LUCERNE.
 Colias hyale.—Buckl., Newm., Sta.
  „ *edusa.*
 Bombyx trifolii.—Sta.
 Strenia clathrata.—Sta.

Medicago lupulina: BLACK MEDICK.—100.
 Colias hyale.—Buckl., Ent. vi. 232. In confinement.
 Zygœna exulans.—Buckl. In confinement.
 Nola centonalis.—Ent. xiii. 43; xv. 41., Buckl.
 Aspilates gilvaria.—Ent. Mag. viii. 116.

Melilotus arvensis : MELILOT.
  Lycæna semiargus (acis).
  Bombyx trifolii.—Sta.
  Euclidia mi.—Newm.

Trifolium pratense : PURPLE CLOVER.—112.
  Colias hyale.—Newm.
  „   *edusa.*—Newm.
  Lycæna icarus (alexis).—Sta.
  „   *corydon.*—Newm.
  Zygæna exulans.—Buckl. In confinement.
  „   *filipendulæ.*—Buckl.
  „   *lonicerœ.*—Buckl., Newm.
  Nola centonalis.—Ent. xiii. 43; xv. 235., Buckl. On the flowers.
  Lithosia griseola.—Buckl.
  „   *caniola.*—Buckl.
  Bombyx trifolii.—Newm., Sta.
  Agrotis saucia.—Newm.
  „   *nigricans.*—Newm.
  Tæniocampa gothica.—Sta.
  Euclidia mi.—Sta.

Trifolium dubium.—100.
  Nola centonalis.—Buckl. On the flowers.

Trifolium ochroleucon.—11.
  Zygæna pilosellæ (minos).—Newm., Sta.
  Calocampa vetusta.

Trifolium repens : DUTCH CLOVER.—112.
  Colias hyale.—Newm.
  „   *edusa.*—Newm.
  Lycæna icarus (alexis).—Sta.
  „   *bellargus (adonis).*
  „   *corydon.*—Newm.
  Zygæna exulans.—Buckl. In confinement.
  „   *filipendulæ.*—Buckl.
  „   *lonicerœ.*—Buckl., Newm
  Nola centonalis.—Ent. xiv. 226., Buckl. On the flowers.
  Lithosia caniola.—Buckl., Newm.
  „   *griseola.*—Buckl.
  Bombyx trifolii.—Newm., Sta.
  Agrotis saucia.—Newm.
  „   *nigricans.*—Newm.
  Tæniocampa gothica.—Sta.
  Euclidia glyphica.—Newm.
  „   *mi.*—Sta.
  Strenia clathrata.—Ent. ix. 179.

Trifolium procumbens : HOP TREFOIL.—101.
    Zygæna trifolii.—Newm., Sta.
      „   *meliloti.*
    Nola centonalis.—Ent. xiii. 43., Buckl. Its proper food.
    Calocampa vetusta.—Newm.

Trifolium filiforme.—61.
    Nola centonalis.—Ent. xiii. 43.

Anthyllis Vulneraria : LADY'S FINGERS ; KIDNEY VETCH.
  —105.
    Lycæna corydon.—Newm., Sta.
      „   *minima (alsus).*—Newm., Buckl., Ent. Mag. iii. 205 ;
        vii. 186 ; x. 43. On the flowers.

Lotus corniculatus : BIRD'S-FOOT TREFOIL.—112.
    Colias hyale.—Newm.
      „   *edusa.*—Buckl.
    Leucophasia sinapis.—Sta.
    Lycæna icarus (alexis).—Sta., Buckl.
      „   *corydon.*—Newm.
    Nisoniades tages.—Buckl., Newm., Sta., Ent. Mag. vi. 233.
    Hesperia comma.—Newm.
    Sesia ichneumoniformis.—Buckl., Ent. Mag. vi. 90. In the
      roots.
    Zygæna pilosellæ (minos).—Newm., Sta.
      „   *exulans.*—Ent. Mag. xx. 150, Buckl. In confine-
        ment.
      „   *meliloti.*
      „   *trifolii.*—Newm., Sta., Buckl.
      „   *loniceræ.*—Buckl., Sta.
      „   *filipendulæ.*—Buckl., Newm.
    Nola centonalis.—Ent. xiii. 43., Buckl.
    Lithosia caniola.—Newm.
    Xylomiges conspicillaris.—Newm., Sta., Ent. Mag. xii. 83.
    Nyssia zonaria.—Ent. xxii. 187.
    Selidosema ericetaria (plumaria).
    Acidalia rubiginata (rubricata). In confinement.
      „   *emutaria.* In confinement.
    Ematurga atomaria.—Newm., Sta.
    Aspilates ochrearia (citraria).—Newm.
    Eubolia plumbaria (palumbaria).
      „   *bipunctaria.*—Newm.

Lotus pilosus : GREATER BIRD'S-FOOT TREFOIL.—100.
    Leucophasia sinapis.—Sta.
    Lycæna argiades.—Ent. xviii. 249.
    Acidalia rubiginata (rubricata). In confinement.
      „   *emutaria.* In confinement.

Lotus pilosus—*continued.*
    Ematurga atomaria.—Newm., Sta.
    Aspilates ochrearia (citraria).—Newm.

Ornithopus perpusillus : BIRD'S-FOOT.—83.
    Lycæna ægon.—Buckl., Newm.
      „   *icarus (alexis).*—Buckl.
    Hesperia comma.—Newm.

Hippocrepis comosa : HORSE-SHOE VETCH.—45.
    Lycæna bellargus (adonis).—Buckl.
      „   *corydon*—Ent. Mag. xi. 113 ; iii. 70 ; iii. 91.
    Zygæna pilosellæ (minos).—Newm., Sta.
      „   *trifolii.*—Newm., Sta.
      „   *loniceræ.*—Sta.

Onobrychis viciæfolia (sativa) : SAINT-FOIN.—27.
    Zygæna filipendulæ.—Buckl.
    Strenia clathrata.—Sta.

Vicia Cracca : TUFTED VETCH.—112.
    Leucophasia sinapis.—Newm., Sta.
    Toxocampa pastinum.—Newm., Sta.

Vicia sylvatica : WOOD VETCH.—72.
    Toxocampa craccæ.—Newm.

Vicia sativa : COMMON VETCH.
    Dasydia obfuscaria (obfuscata).—Newm., Sta.
    Acidalia marginepunctata (promutata)—Sta.

Lathyrus (Orobus) pratensis : MEADOW VETCHLING.—109.
    Zygæna loniceræ.—Buckl.

Lathyrus (Orobus) tuberosus : TUBEROUS BITTER VETCH.
    Leucophasia sinapis.—Newm., Sta.

Lathyrus (Orobus) sylvestris : EVERLASTING PEA.—60.
    Leucophasia sinapis.—Sta.
    Polia flavicincta.—Ent. Mag. ix. 248.

Colutea arborescens : BLADDER SENNA. (NOT BRITISH).
    Lycæna bætica.—Newm., Ent. xii. 83. In the pods.

Robinia Pseud-acacia : ACACIA.
    Orgyia antiqua.—Buckl.
    Amphidasys betularia.—Newm.

Prunus spinosa : BLACKTHORN.—103.
    Aporia cratægi.—Sta.
    Thecla pruni.—Newm., Sta.
      „   *betulæ.*—Buckl., Newm., Sta.
    Smerinthus ocellatus.

Prunus spinosa—*continued.*
   Nola cucullatella.—Newm., Sta.
   Porthesia chrysorrhœa.—Newm., Sta., Ent. Mag. viii. 18.
   Ocneria dispar.—Newm.
   Orgyia gonostigma.
      „ *antiqua.*
   Trichiura cratægi.—Newm., Sta.
   Eriogaster lanestris.—Sta.
   Lasiocampa quercifolia.—Newm., Sta.
   Saturnia pavonia (carpini).—Newm.
   Cilix glaucata (spinula).—Sta
   Demas coryli.—Sta.
   Acronycta tridens.—Newm.
      „ *psi.*
      „ *strigosa.*—Sta.
   Noctua augur.—Ent. xi. 269.
   Tryphœna ianthina.—Ent. xi. 269.
   Tæniocampa incerta (instabilis).—Sta.
   Orthosia lota.—Ent. xxii. 151.
   Cerastis spadicea.—Newm., Sta.
   Valeria oleagina.—Newm., Sta.
   Miselia oxyacanthæ.—Newm., Sta.
   Xylomiges conspicillaris.—Ent. xvi. 132.
   Uropteryx sambucaria (sambucata).—Ent. xiii. 140.
   Rumia luteolata (cratægata).—Newm., Sta.
   Angerona prunaria.—Newm.
   Selenia lunaria.—Newm., Sta.
   Crocallis elinguaria.—Newm., Sta.
   Boarmia repandata.—Sta.
   Tephrosia crepuscularia.—Ent. xix. 159.
   Asthena sylvata.—Ent. xii. 296.
   Bapta temerata.—Newm., Sta.
   Aleucis pictaria.—Newm.
   Abraxas grossulariata.—Newm., Sta.
   Hybernia rupicapraria.—Newm., Sta.
      „ *defoliaria.*—Newm., Sta.
   Anisopteryx æscularia.—Newm., Sta.
   Cheimatobia brumata.
   Oporabia dilutata.—Newm.
   Melanthia bicolorata (rubiginata).—Newm.
   Cidaria corylata.—Newm., Sta.

Prunus : PLUM.
   Aporia cratægi.—Sta.
   Nola cucullatella.—Newm.
   Zeuzera pyrina (æsculi).—Newm.   In the wood.
   Ocneria dispar.—Newm.
   Dasychira fascelina.—Newm.

Prunus—*continued.*
   Orgyia antiqua.
   Bombyx neustria.
   Phalera bucephala.
   Acronycta psi.
   Diloba cæruleocephala.
   Mania typica.—Newm. Before hybernation.
   Tæniocampa munda.—Newm.
   Dasychampa rubiginea.—Ent. Mag. v. 206. In confinement.
   Calymnia pyralina.—Newm., Sta.
   Catocala nupta.—Ent. xv. 133
   Angerona prunaria.—Sta.
   Selenia bilunaria (illunaria).—Sta.
   Eugonia (Ennomos) autumnaria (alniaria).—Newm. In confinement.
   Biston hirtaria.—Newm.
   Boarmia repandata.—Newm.
      „   *gemmaria (rhomboidaria).*—Newm., Sta.
   Tephrosia biundularia.—Newm. In confinement.
   Acidalia rubiginata (rubricata).

Prunus Avium : WILD CHERRY.—90.
   Vanessa polychloros.—Newm.
   Acronycta alni.—Ent. Mag. xx. 82.
   Bapta temerata.—Newm., Sta.
      „   *taminata.*—Newm.

Spiræa Filipendula : DROPWORT.—61.
   Xanthia aurago.—Ent. xii. 297.

Spiræa Ulmaria : MEADOW-SWEET.—112.
   Callimorpha dominula.—Ent. xiii. 172.
   Saturnia pavonia (carpini).—Ent. xiii. 172, Ent. Mag. xvii. 124.
   Tæniocampa gracilis.—Ent. xvi. 199.
   Anchocelis litura.—Newm.
   Acidalia marginepunctata (promutata).—Ent. xiv. 212. (?).
      „   *aversata.*—Newm.
      „   *immutata.*—Ent. xiv. 212.

Rubus fruticosus : BRAMBLE
   Thecla rubi.—Newm., Sta.
   Syrichthus malvæ (alveolus).—Buckl., Newm., Ent. Mag. xi 236.
   Bombyx rubi.—Newm.
   Thyatira derasa.—Newm., Sta.
      „   *batis.*—Newm., Sta.
   Acronycta auricoma.—Newm., Sta., Ent. Mag. iii. 261.
   Noctua triangulum.
      „   *dahlii.* In confinement.

Rubus fruticosus—*continued.*
   Anchocelis litura.—Ent. Mag. ix. 39.
   Hadena rectilinea.—Sta.
   Aplecta occulta.—Ent. Mag. xii. 66.   In confinement.
   Erastria venustula—Ent. xvi. 164.   On the flowers.
     ,, *fasciana (fuscula).*—Newm., Sta.
   Uropteryx sambucaria (sambucata).—Sta.
   Boarmia repandata.—Sta., Ent. xi. 268.
   Nemoria viridata.—Sta.
   Acidalia bisetata.—Ent. Mag. v. 96.   In confinement, on withered leaves.
   Acidalia degeneraria.—Ent. Mag. ix. 115.   In confinement.
     ,, *rusticata.*—Ent. Mag. iii. 259.   In confinement, on withered leaves.
   Melanthia albicillata.—Newm., Sta.

Rubus Idæus: RASPBERRY.—105.
   Syrichthus malvæ (alveolus).—Newm., Sta.
   Acronycta alni.—Ent. Mag. xx. 82.
   Erastria fasciana (fuscula).—Newm.
   Zanclognatha tarsipennalis.—Sta.
   Melanthia albicillata.—Newm., Sta.

Rubus cæsius: DEWBERRY.—68.
   Nola albulalis.—Buckl., Ent. ix. 177.

Geum urbanum: COMMON AVENS.—104.
   Acidalia aversata.—Newm., Sta.

Geum rivale: WATER AVENS.—90.
   Acidalia aversata.—Newm.

Fragaria vesca.—WILD STRAWBERRY.—111.
   Acronycta rumicis.—Newm.
   Noctua flammatra.
   Mania maura.—Newm.
   Gnophos obscuraria (obscurata).—Ent. xiii. 13.
   Melanthia albicillata.—Newm.   In confinement.
   Cidaria truncata (russata).—Newm.
     ,, *immanata.*—Newm.

Potentilla Fragariastrum: STRAWBERRY-LEAVED CINQUEFOIL.—101.
   Syrichthus malvæ (alveolus).—Buckl.
   Epunda lutulenta.—Ent. Mag. vi. 235.   In confinement.

Potentilla (Sibbaldia) procumbens.—SCOTCH CINQUEFOIL.—69
   Zygæna exulans.   Buckl.

Potentilla reptans: TRAILING TORMENTIL—89.
   Tryphæna orbona (subsequa).—Ent. Mag. ix. 57.
   Erastria venustula.—Newm. On the flowers.
   Aspilates gilvaria.—Ent. Mag. viii. 116.

Potentilla Anserina: SILVER-WEED.—112.
   Nola centonalis.—Ent. xiii. 43.
   Erastria venustula.—Ent. xvi. 164. On the flowers.

Alchemilla alpina: ALPINE LADY'S MANTLE.—27.
   Zygæna exulans.—Ent. Mag. xx. 151., Buckl.

Poterium Sanguisorba: COMMON SALAD BURNET.—70.
   Ino globulariæ.—Ent. xv. 189.
   Acosmetia caliginosa.
   Agrotis ashworthii.—Newm.
   Gnophos obscuraria.—Newm., Sta.

Rosa spinosissima: BURNET ROSE.—90.
   Bombyx rubi.—Ent. Mag. xxiii. 60.
   Tæniocampa opima.—Ent. Mag. ix. 21.
   Anticlea badiata.—Ent. xiv. 87.
   Cidaria fulvata.

Rosa canina: DOG ROSE.—110.
   Acronycta alni.—Ent. Mag. xx. 82.
   Hemerophila abruptaria.—(?). Also on cultivated rose.
   Boarmia gemmaria (rhomboidaria).—(?). Also on cultivated rose.
   Anticlea badiata.—Newm., Sta.
      „ *nigrofasciaria (derivata).*—Newm., Sta.
   Cidaria sitterata (psittacata).—Sta.
      „ *fulvata.*—Newm., Sta.

Pyrus Aria: WHITE-BEAM.—46.
   Vanessa polychlorus.—Newm.

Pyrus Aucuparia: MOUNTAIN ASH.—106.
   Acronycta alni.—Ent. xii. 251.
   Venusia cambrica (cambricaria).—Newm.
   Lobophora viretata.—Ent. xx. 183.

Pyrus communis: PEAR.—49.
   Aporia cratægi.—Sta.
   Vanessa polychlorus.—Newm.
   Sesia myopiformis.—Newm. In the wood.
   Zeuzera pyrina (æsculi).—Newm., Sta. In the wood.
   Orgyia antiqua.
   Bombyx neustria.—Sta.
   Cilix glaucata (spinula).
   Diloba cæruleocephala.

Pyrus communis—continued.
    Acronycta psi—Newm.
      „   *tridens.*
    Cosmia pyralina.—Newm., Sta.
    Mania typica.—Newm.
    Crocallis elinguaria.—Newm.
    Eugonia (Ennomos) autumnaria (alniaria).—Newm.
    Biston hirtaria.—Newm.
    Amphidasys betularia.—Ent. xix. 253.
    Cheimatobia brumata.—Newm.
    Eupithecia consignata.—Newm.
      „   *rectangulata.*—Newm., Sta. In the buds.

Pyrus Malus: APPLE.
    Aporia cratægi.—Ent. xviii. 179., Sta.
    Chærocampa elpenor.—Ent. xviii. 181.
    Smerinthus ocellatus.—Newm., Sta., Buckl.
    Sesia myopiformis.—Buckl., Newm., Sta. In the bark and wood.
    Arctia caia (caja).—Ent. xvi. 162.
    Zeuzera pyrina (æsculi).—Newm., Sta. In the wood.
    Porthesia similis (auriflua).—Sta.
    Ocneria dispar.—Newm.
    Psilura monacha.—Sta.
    Orgyia antiqua.
    Bombyx neustria.—Newm., Sta.
    Stauropus fagi.—Ent. xv. 89.
    Acronycta psi.
      „   *alni.*—Ent. x. 254.
    Dasycampa rubiginea.—Newm.
    Rumia luteolata (cratægata).—Ent. xvi. 162.
    Crocallis elinguaria.—Newm.
    Eugonia (Ennomos) autumnaria (alniaria).—Newm.
    Amphidasys betularia.—Ent. xix. 253.
    Cheimatobia brumata.—Newm.
    Eupithecia consignata.—Newm., Sta.
      „   *rectangulata.*—Newm., Sta.
    Cidaria siterata (psittacata).—Sta.

Cratægus Oxyacantha: WHITETHORN.—110.
    Aporia cratægi.—Newm., Sta.
    Nola cucullatella.—Sta., Ent. xii. 164.
    Spilosoma lubricepeda.—Buckl.
    Porthesia (liparis) chrysorrhœa.—Newm., Sta., Ent. Mag. viii. 18.
      „   „   *similis (auriflua).*—Newm., Sta.
    Ocneria dispar.—Newm., Ent. xviii. 263.
    Orgyia gonostigma.
      „   *antiqua.*

Cratægus Oxyacantha—*continued.*
   Zeuzera pyrina (æsculi).—Ent. xx. 137.
   Trichiura cratægi.—Newm., Sta.
   Pœcilocampa populi.—Sta.
   Eriogaster lanestris.—Newm., Sta.
   Bombyx quercus.—Buckl., Newm.
       „   „ *v. callunæ.*—Buckl., Ent. Mag. vii. 88.
   Saturnia pavonia (carpini).— Ent. ix. 186, Ent. Mag. vii. 88.
   Cilix glaucata (spinula).—Newm., Sta.
   Acronycta psi.—Newm.
       „   *tridens.*—Newm.
       „   *alni.*—Newm.
       „   *strigosa.*—Newm.
       „   *menyanthidis.*— Ent. Mag. vii. 88. In confinement.
   Diloba cæruleocephala.—Buckl., Newm., Sta.
   Noctua augur.—Newm., Ent. xi. 269. After hybernation.
       „   *triangulum.* In confinement.
       „   *dahlii.* In confinement.
       „   *sobrina.* In confinement.
       „   *xanthographa.*
   Tryphæna comes (orbona).—Newm. After hybernation.
       „   *ianthina.*—Ent. xi. 269.
   Tæniocampa gothica.—Newm.
       „   *miniosa.*—Newm.
   Cerastis spadicea.—Sta.
   Polia chi.—Newm.
   Miselia oxyacanthæ.—Newm., Sta.
   Aplecta nebulosa.—Newm.
   Calocampa solidaginis.—Ent. Mag. ix. 92.
   Amphipyra tragopogonis.—Newm.
   Uropteryx sambucaria (sambucata).—Ent. xiii. 40.
   Rumia luteolata (cratægata) —Newm., Sta.
   Crocallis elinguaria.—Sta.
   Eugonia (Eunomos) quercinaria (angularia).—Ent. ix. 50.
   Boarmia repandata.—Ent. xvi. 270.
   Nemoria viridata.—Sta.
   Hemithea strigata (thymiaria).—Sta.
   Acidalia contiguaria.—Ent. Mag. iii. 69.
       „   *trigeminata.* In confinement.
       „   *immutata.* In confinement.
       „   *degenaria.* In confinement.
   Hybernia rupicapraria.—Newm., Sta.
       „   *aurantiaria.*—Newm.
       „   *defoliaria.*—Newm., Sta.
   Anisopteryx æscularia.—Newm., Sta.
   Cheimatobia brumata.
   Oporabia dilutata.—Newm.

Crataegus Oxyacantha—*continued.*
  Eupithecia vulgata.—Newm.
  „   *dodoneata.*—(?).
  „   *exiguata.*—Newm., Sta.
  Coremia quadrifasciaria.—Newm., Sta.
  Cidaria truncata (russata).—Newm.
  „   *dotata (pyraliata).*—Sta.

Saxifraga aizoides.—32.
  Larentia flavicinctata.—Ent. Mag. xii. 86.
  „   *cæsiata.*—Ent. Mag. xii. 86.

Saxifraga granulata : WHITE MEADOW SAXIFRAGE —75.
  Larentia flavicinctata.—Newm., Sta.

Saxifraga hypnoides : MOSSY SAXIFRAGE.—46.
  Larentia flavicinctata.—Sta., Ent. Mag. xii. 5 ; xii. 68.

Chrysosplenium alternifolium : ALTERNATE-LEAVED GOLDEN SAXIFRAGE.—67.
  Zanclognatha grisealis (nemoralis).—Sta.

Ribes Grossularia : GOOSEBERRY.
  Halia vauaria (wavaria).—Newm., Sta.
  Abraxas grossulariata.—Newm.
  Cidaria prunata.—Newm., Sta.

Ribes rubrum : RED CURRANT.
  Vanessa c-album.—Newm , Sta.
  Sesia tipuliformis.—Buckl., Newm., Sta.   In the shoots.
  Halia vauaria (wavaria).—Sta.
  Abraxas grossulariata.—Sta.
  Cidaria prunata.—Newm., Sta.
  „   *associata (dotata).*—Newm., Sta.

Ribes nigrum : BLACK CURRANT.
  Sesia tipuliformis.—Buckl , Newm., Sta.   In the shoots.
  Eupithecia assimilata.—Newm., Sta.
  „   *exiguata.*—Newm., Sta.
  Cidaria prunata.—Newm., Sta.
  „   *associata (dotata).*—Newm., Sta.

Ribes sanguinem : FLOWERING CURRANT (NOT BRITISH).
  Abraxas grossulariata.—Ent. xix. 255.

Cotyledon Umbilicus : WALL PENNY-WORT.—53.
  Abraxas grossulariata.—Ent. xiv. 18.

Sedum Telephium : ORPINE.—75.
  Caradrina morpheus —Ent. Mag. x. 254.
  Abraxas grossulariata.—Ent. xiv. 43.

Sedum acre : BITING STONECROP.—104.
  Agrotis lucernea.—Newm.

Epilobium angustifolium : FLOWERING WILLOW.—92.
  Chærocampa porcellus.—Ent. xii. 250.

Epilobium hirsutum : GREAT HAIRY WILLOW-HERB.—93.
  Chærocampa celerio.—Ent. xix. 124.
  „   elpenor.—Buckl., Newm., Sta.
  „   porcellus.—Ent. xii. 250.
  Spilosoma urticæ.—Newm.
  Mania typica.—Sta.
  Cidaria silaceata.—Newm. Also on other Epilobia.

Lythrum Salicaria : PURPLE LOOSESTRIFE.—92.
  Chærocampa elpenor.
  „   porcellus.—Ent. xii. 250.
  Saturnia pavonia (carpini).
  Acidalia immutata.

Circæa lutetiana : ENCHANTER'S NIGHTSHADE.—97.
  Chærocampa elpenor.—Ent. xv. 234.
  Cidaria silaceata.—Newm.

Pimpinella Saxifraga : COMMON BURNET SAXIFRAGE.—99.
  Zygæna pilosellæ (minos) v. nubigena.—Buckl.
  Acidalia dimidiata (scutulata).—Newm. On the flowers.
  Eupithecia oblongata (centaureata).—Newm. On the flowers.
  „   virgaureata.—Sta. On the flowers.
  „   pimpinellata.—Newm., Sta. On the flowers.

Pimpinella major : GREAT BURNET SAXIFRAGE.—50.
  Eupithecia oblongata (centaureata).—Newm. On the flowers.
  „   virgaureata.—(?). On the flowers.

Conopodium (Bunium) denudatum : COMMON PIG-NUT. 104.
  Tanagra atrata (chærophyllata).—Newm. On the blossoms.

Chærophyllum temulum : ROUGH CHERVIL.—97.
  Larentia didymata.—Sta.
  Tanagra atrata (chærophyllata).—Sta.

Anthriscus sylvestris : HARE'S PARSLEY.—102.
  Plusia bractea.—Ent. xv. 21.
  Acidalia dimidiata (scutulata).—Newm. On the flowers.
  Larentia didymata.—Newm.
  Eupithecia pumilata.—Newm. On the flowers.

Silaus pratensis : MEADOW SAXIFRAGE.—65.
  Eupithecia oblongata (centaureata).—Newm. On the flowers.

Angelica sylvestris : WILD ANGELICA.—110.
    Papilio machaon.—Ent. Mag. xviii. 245., Buckl.
    Eupithecia albipunctata.—Newm. On the flowers.
      „   *trisignaria.*—Newm., Ent. ix. 260. On the flowers.
      „   *coronata.*—Newm. On the flowers.

Peucedanum palustre : MILK PARSLEY.—12.
    Papilio machaon.—Newm., Sta.

Heracleum sphondylium : COW PARSNEP.—110.
    Papilio machaon.—Newm.
    Dasypolia templi.—Newm.
    Eupithecia albipunctata.—Newm. On the flowers.

Daucus Carota : WILD CARROT.—106.
    Chærocampa celerio.
    Bombyx castrensis.—Newm., Sta.
    Aspilates ochearia (citraria).—Newm.
    Eupithecia scabiosata (subumbrata).—Ent. xviii. 264.

Hedera Helix : IVY.—111.
    Lycæna argiolus.—Buckl., Newm., Sta. On the blossoms.
    Polia flavicincta (flavocincta).—Ent. xix. 128.
    Uropteryx sambucaria (sambucata).—Newm., Sta.
    Boarmia gemmaria (rhomboidaria).—Ent. xxi. 278 ; Ent. Mag. iii. 20.
    Acidalia rusticata. In confinement.
    Lobophora viretata.—Ent. x. 98.

Cornus sanguinea : DOG-WOOD.—66.
    Lycæna argiolus.—Buckl.
    Sesia andreniformis.—Ent. xx. 102., Ent. Mag. x. 161.
    Lithosia lurideola (complanula).—Buckl.

Sambucus nigra : ELDER.—106.
    Saturnia pavonia (carpini).—Ent. xxii. 17.
    Gortyna ochracea (flavago).—Newm. In the stems.
    Mamestra persicariæ.—Newm.
    Uropteryx sambucaria (sambucata).—Newm., Sta.
    Pericallia syringaria.—Newm.
    Eugonia (Ennomos) autumnaria (alniara).—Newm.

Viburnum Opulus : GUELDER ROSE.—75.
    Lobophora viretata.—Ent. x. 98.

Lonicera Periclymenum : HONEYSUCKLE.—109.
    Melitæa aurinia (artemis).—Ent. xvii. 258.
    Limenitis sibylla.—Newm., Sta.
    Macroglossa fuciformis.—Buckl., Sta., Ent. Mag. v. 226.
    Cerastis spadicea.—Newm.

Lonicera Periclymenum—*continued.*
Polia flavicincta (flavocincta).—Ent. xix. 92.
Hadena porphyrea (satura).—Sta.
„ *thalassina.*—Sta.
Xylocampa areola (lithoriza).—Newm., Sta.
Plusia iota.—Sta.
Uropteryx sambucaria (sambucata).—Newm.
Crocallis elinguaria.—Newm.
Acidalia degenaria. In confinement.
Lobophora carpinata (lobulata).—Newm.
„ *polycommata.*—Newm., Sta.
Anticlea nigrofasciaria (derivata).—Sta.

Rubia peregrina: MADDER.—23.
Deilephila galii.
Macroglossa stellatarum.

Galium verum: LADY'S BEDSTRAW.—109.
Deilephila galii.—Newm., Sta.
„ *livornica.*—Newm., Sta.
Chærocampa porcellus.—Buckl., Newm., Sta.
„ *elpenor.*—Ent. xviii. 288., Newm.
Macroglossa stellatarum.—Buckl., Newm.
„ *fuciformis.*—Newm.
Agrotis aquilina.—Sta.
„ *obelisca.*—Newm.
Noctua plecta.—Newm.
Acidalia ochrata.—Ent. xiii. 306. On the flowers.
„ *perochraria (ochrearia).* On the flowers. In confinement.
„ *emarginata.*—Sta.
Larentia multistrigaria.—Newm., Sta. Also on other *Galia.*
„ *salicata.*—Newm. Also on other *Galia.*
Melanthia ocellata.—Newm., Sta.
Melanippe galiata.—Newm., Sta.
Anticlea cucullata (sinuata).—Newm., Sta., Ent. ix. 232.
„ *rubidata.*—Newm., Sta.
Coremia unidentaria.—Newm.
Phibalapteryx vittata (lignata).—Ent. Mag. viii. 18; xxi. 158.
Mesotype virgata (lineolata).—Newm., Sta., Ent. Mag. ix. 197; x. 255.

Galium Mollugo: HEDGE BEDSTRAW.—75.
Deilephila galii.
Chærocampa celerio.—Ent. xix. 124.
„ *porcellus.*—Buckl.
„ *elpenor.*—Buckl. In confinement on the flowers.
Macroglossa stellatarum.—Buckl., Sta.

Galium Mollugo—*continued.*
    Caradrina morpheus.—Ent. Mag. x. 254.
    Noctua umbrosa.—Ent. Mag. viii. 140. In confinement.
    Epunda nigra.—Newm.
    Acidalia dimidiata (scutulata).—Ent. ix. 12.
    Larentia olivata.—Newm., Ent. Mag. xi. 86.
        „ *viridaria (pectinitaria).*—Newm.
    Eupithecia scabiosata (subumbrata).—Newm.
        „ *satyrata.*—Newm., Sta.
    Melanippe tristata.—Newm., Sta.
        „ *rivata.*—Newm., Sta.
        „ *sociata (subtristata).*—Newm., Sta.
        „ *galiata.*—Newm., Sta.
    Anticlea rubidata.—Newm., Sta.
    Phibalapteryx vittata (lignata).—Ent. Mag. xxi. 158.
    Cidaria suffumata.—Newm.
        „ *dotata (pyraliata).*—Newm.
        „ *picata.*—Ent. ix. 13.

Galium sexatile : HEATH BEDSTRAW.—111.
    Larentia viridaria (pectinitaria).—Newm.
    Phibalapteryx vittata (lignata).—Ent. Mag. viii. 18. In confinement.
    Mesotype virgata (lineolata).—Ent. Mag. ix. 197 ; x. 255. In confinement.

Galium palustre : WATER BEDSTRAW.—112.
    Chærocampa porcellus.—Buckl.
        „ *elpenor.*—Buckl.
    Acidalia perochraria (ochrearia). On the flowers. In confinement.
    Phibalapteryx vittata (lignata).—Ent. Mag. viii. 18.

Galium Aparine : GOOSE GRASS.—112.
    Macroglossa stellatarum.
    Cidaria dotata (pyraliata).—Newm.

Asperula odorata : SWEET WOODRUFF.—101.
    Noctua plecta.—Newm.
    Larentia salicata.—Newm. In confinement.
        „ *multistrigaria.*—Newm. In confinement.
    Coremia unidentaria.—Newm.

Valeriana officinalis : GREAT WILD VALERIAN.—111.
    Zygæna filipendulæ.
    Acidalia marginepunctata (promutata).—Ent. xiv. 212.
        „ *immutata.*—Ent. xiv. 212.
    Eupithecia valerianata (viminata).—Newm., Ent. iii. 45, 19. On the flowers.

Dipsacus sylvestris: WILD TEASEL.—72.
    Caradrina morpheus.—Sta.
Dipsacus pilosus: SMALL TEASEL.—51.
    Caradrina morpheus.—Sta.
Scabiosa succisa: DEVIL'S-BIT SCABIOUS.—112.
    Melitæa aurinia (artemis).—Newm., Sta.
    Macroglossa bombyliformis.—Ent. xiii. 123 ; xx. 99., Buckl.
    * *Syntomis phegea.*
    Agrotis ashworthii.—Newm.
    Calocampa vetusta.—Newm.
        „   *exoleta.*—Newm.
    Eupithecia satyrata.—Sta. On the flowers.
        „   *minutata.*—Ent. xvi. 62. On the flowers.

Scabiosa arvensis: FIELD SCABIOUS.—95.
    Macroglossa bombyliformis.—Newm., Sta.
        „   *fuciformis.*—Newm.
    Epunda lutulenta.—Ent. Mag. vi. 236.
    Calocampa vetusta.—Newm.
    Eupithecia scabiosata (subumbrata).—Newm. On the flowers.
        „   *satyrata.*—Newm. On the flowers.

Scabiosa Columbaria: SMALL SCABIOUS.—71.
    Calocampa vetusta.—Newm.
    Eupithecia oblongata (centaureata).—Newm. On the flowers.

Eupatorium cannabinum: HEMP AGRIMONY.—97.
    Gortyna ochracea (flavago).—Newm. In the stems.
    Plusia chryson (orichalcea).—Newm., Sta.
        „   *bractea.*—Newm.
    Eupithecia oblongata (centaureata).—Newm. On the flowers.
        „   *absynthiata.*—Newm., Ent. ix. 260. On the flowers.
        „   *coronata.*—Newm. On the flowers.

Solidago Virgaurea: GOLDEN-ROD.—108.
    Agrotis ashworthii.—Newm., Ent. xxiii. 6.
    Cucullia asteris.—Newm., Sta.
        „   *gnaphalii.*—Newm., Sta.
    Acidalia ochrata.—Ent. xiii. 307.
    Eupithecia oblongata (centaureata).—Newm.
        „   *pernotata.*—Ent. xiii. 146., Newm., Sta.
        „   *virgaureata.*—Newm.
        „   *vulgata.*—Sta.
        „   *expallidata.*—Newm., Ent. Mag. x. 118.
        „   *absinthiata.*—Newm., Sta.
        „   *coronata.*—Newm.

Aster tripolium: SEA STAR-WORT.—65.
    Cucullia asteris.—Newm.

Filago germanica : COMMON CUD-WEED.—93.
  Vanessa cardui.—Ent. xix. 132.
Achillea Millefolium : YARROW ; MILFOIL.—112.
  Nyssia zonaria.—Newm., Sta.
  Acidalia marginepunctata (*promutata*).—Sta.
  „   *immutata.*
  Aspilates gilvaria.—Newm., Sta., Ent. Mag. viii. 116.
  Eupithecia oblongata (*centaureata*).—Newm. On the flowers.
   „   *succentauriata.*—Sta. On the seeds.
   „   *subfulvata.*—Ent. xviii. 144, Newm. On the
        flowers.
   „   *absinthiata.*—Newm. On the flowers.
Anthemis Cotula : 72.
  Cucullia chamomillæ.—Newm., Sta.
  Sterrha sacraria.—Newm.
Anthemis nobilis : CHAMOMILE.—46.
  Cucullia chamomillæ.—Sta.
  Sterrha sacraria.—Newm.
  Eupithecia succentauriata.—Ent. Mag. x. 118.
Anthemis arvensis : CORN CHAMOMILE.—69.
  Cucullia chamomillæ.—Sta.
  Sterrha sacraria.—Newm.
Chrysanthemum Parthenium : COMMON FEVERFEW.
  Cucullia chamomillæ.
Matricaria inodora : CORN FEVERFEW.—111.
  Cucullia chamomillæ.—Sta., Newm.
  Heliothis peltigera.—Newm.
Matricaria Chamomilla : DOG'S CHAMOMILE.—62.
  Cucullia chamomillæ.—Sta.
Tanacetum vulgare : TANSY.—102.
  Eupithecia absinthiata.—Ent. xiv. 222. On the flowers.
Artemisia Absinthium : COMMON WORMWOOD.—71.
  Cucullia absinthii.—Newm., Sta.
   „   *artemisiæ.*—Ent. xviii. 290.
  Eupithecia absinthiata.—Sta. On the flowers.
   „   *extensaria.*—Ent. xviii. 146 (?). On the flowers.
   „   *innotata.*—Ent xviii. 146 (?). On the flowers.
   „   *succentauriata.*—Sta. On the seeds.
Artemisia vulgaris : MUGWORT.—110.
  Heliothis scutosa.—Ent. x. 106.
  * *Emydia striata* (*grammica*).—Newm. On the seeds.
  Eupithecia succentauriata.—Newm.
   „   *absinthiata.*—Ent. xviii. 144., Newm., Sta. On the
        flowers.

Artemisia campestris : FIELD WORMWOOD.—4.
   Agrotis vestigialis (*valligera*).
   Heliothis scutosa.—Sta., Ent. x. 106.

Artemisia maritima ; SEA WORMWOOD.—47.
   Bombyx castrensis.—Newm., Sta.
   Phorodesma smaragdaria.—Ent. xxi. 190.
   Mamestra abjecta.

Tussilago Farfara : COLT'S-FOOT.—112.
   Acidalia ochrata.—Ent. xiv. 159. On the flowers.

Petasites officinalis (vulgaris) : BUTTER-BUR.—102.
   Hydrœcia petasitis.—Newm., Sta. In the stems and roots.

Senecio vulgaris : GROUNDSEL.—112.
   Euchelia jacobææ.—Newm.
   Nemeophila plantaginis.—Ent. xvi. 113.
   Polia flavicincta (*flavocincta*).—Newm.
   Phlogophora meticulosa.—Newm.
   Plusia bractea.—Ent. xv. 21 ; xvi. 134.
     ,,   *iota.*—Sta.
     ,,   *pulchrina.*—Sta.
   Acidalia immutata. In confinement.
   Eupithecia oblongata (centaureata).—Sta. On the flowers.
     ,,   *absinthiata.*—Sta. On the flowers.
   Coremia munitata.—Newm. In confinement.
   Camptogramma fluviata.—Newm. In confinement.

Senecio erucifolius : HOARY RAGWORT.—67.
   Eupithecia oblongata (centaureata).—Newm. On the flowers.
     ,,   *absinthiata.*—Newm. On the flowers.

Senecio Jacobæa : RAGWORT.—112.
   Polyommatus phlæas.—Newm.
   Euchelia jacobææ.—Sta.
   Tæniocampa opima.—Ent. Mag. ix. 21.
   Epunda lichenea.—Newm., Sta.
   Eupithecia oblongata (centaureata).—Newm. On the flowers.
     ,,   *succentauriata.*—Ent. Mag. xxiii. 61. On the flowers.
     ,,   *virgaureata.*—Ent. ix. 260. On the flowers.
     ,,   *expallidata.*—Newm (?). On the flowers.
     ,,   *absinthiata.*—Newm., Ent. ix. 260. On the flowers.
   Eupithecia vulgata.—(?).
     ,,   *castigata.*—(?).
     ,,   *virgaureata.* In confinement.

Senecio palustris : MARSH FLEA-WORT.—7.
   Eupithecia virgaureata. In confinement.

Arctium minus (Lappa): BURDOCK.—38.
    Vanessa cardui.
    Hepialus humuli.—Newm., Sta. On the roots.
    Gortyna ochracea (flavago).—Newm., Sta. In the stems.
    Plusia chrysitis.—Newm.

Carduus nutans: MUSK THISTLE.—73.
    Vanessa cardui.—Sta.
    Gortyna ochracea (flavago).—Newm., Sta. In the stems.
    Agrotis obscura (ravida).—Newm. On the roots.

Carduus crispus: WELTED THISTLE.—87.
    Vanessa cardui.—Sta.

Cnicus lanceolatus: SPEAR THISTLE.—112.
    Vanessa cardui.—Sta.

Cnicus palustris: MARSH THISTLE.—112.
    Gortyna ochracea (flavago).—Newm. In the stems.

Cnicus arvensis: FIELD THISTLE.—112.
    Vanessa cardui.—Newm.
    Calocampa exoleta.—Newm.

Onopordon Acanthium: SCOTCH THISTLE.—60.
    Vanessa cardui.—Buckl.

Centaurea nigra: KNAP-WEED.—110.
    Sesia ichneumoniformis.—Ent. xvi. 128.
    Ino globulariæ.—Ent. Mag xx. 97., Buckl.
    Nyssia zonaria.—Ent. xix. 158.
    Ematurga atomaria.—Sta.
    Eupithecia oblongata (centaureata).—Sta. On the flowers.
        "   *scabiosata (subumbrata).*—Newm. On the flowers.
        "   *satyrata.*—Newm. On the flowers.

Cichorium Intybus: CHICORY.—63.
    Agrotis suffusa.
    Aporophyla australis.—Newm.

Crepis virens: SMOOTH HAWK'S-BEARD.—106.
    Heliothis dipsacea.—Ent. Mag. xi. 257. On the flowers and seeds.
    Acidalia ochrata.—Ent. xiii. 307. On the flowers.

Hieracium Pilosella: MOUSE-EAR HAWK-WEED.—110.
    Nemeophila russula.—Newm.
    Agrotis ashworthii.—Newm. Also on other *Hieracia.*
    Hecatera serena.—Sta. Also on other *Hieracia.*
    Plusia bractea.—Ent. xiv. 222.

Leontodon hispidus: ROUGH HAWK-BIT.—87.
    Eupithecia scabiosata (subumbrata).—Newm. On the flowers.
        "   *satyrata.*—Newm. On the flowers.

Taraxacum officinale : DANDELION.—110.
  * *Syntomis phegea.*
  Hepialus hectus.—Sta. On the roots.
  „ *humuli.*—Buckl. On the roots.
  Nemeophila russula.—Newm., Sta.
  Callimorpha hera.
  Noctua flammatra.
  Agrotis obscura (ravida). On the roots and leaves.
  „ *lunigera.*—Ent. Mag. iii. 188. In confinement.
  „ *lucernea.*—Sta.
  „ *puta.*—Newm. (?).
  Dasycampa rubiginea.—Newm.
  Aplecta occulta.—Sta.
  Hadena dentina.—Newm., Sta. On the roots.
  Acidalia ochrata.—Ent. xiv. 159.
  „ *bisetata.*—Sta.
  „ *dilutaria (osseata).*—Ent. xi. 19., Ent. Mag. v. 26. In confinement.
  „ *virgularia.*—Ent. xi. 91. In confinement.
  „ *strigilaria.*—Ent. Mag. viii. 91. In confinement.
  „ *immutata.* In confinement.
  „ *scutulata.*—Ent. Mag. v. 96. In confinement.

Lactuca virosa : SLEEPWORT.—51.
  Hecatera chrysozona (dysodea). On the flowers and seeds. Also on other *Lactucæ.*
  „ *serena.*—Newm.
  Polia chi.—(?).
  Cucullia umbratica.—Newm.

Sonchus oleraceus : SOW-THISTLE.—109.
  Hecatera serena.—Newm., Sta.
  Polia chi.—Sta.
  Cucullia umbratica.—Newm., Sta.

Sonchus arvensis : CORN SOW-THISTLE.—107.
  Hecatera serena.—Newm., Sta.
  Polia chi.—Sta.
  Cucullia umbratica.—Newm., Sta.

Sonchus palustris : MARSH SOW-THISTLE.—7.
  Cucullia umbratica.—Newm.

Jasione montana : SHEEP'S SCABIOUS.—78.
  Eupithecia jasioneata.

Campanula glomerata : CLUSTERED BELL-FLOWER.—49.
  Eupithecia oblongata (centaureata).—Newm. On the flowers.

Campanula Trachelium : CANTERBURY BELLS.—58.
  Eupithecia campanulata (denotata).—Newm., Ent. Mag. vii. 143. On the seeds. Also on most *Campanulæ.*

Campanula rotundifolia : HAREBELL.—110.
  Agrotis lucernea.—Newm. In confinement.
  „   *ashworthii.*—Sta. In confinement.
  Polia xanthomista.—Newm.

Vaccinium Vitis-Idæa : COW-BERRY.—62.
  Hadena rectilinea.—Sta.
  Anarta cordigera.—Newm., Sta.
  Larentia cæsiata.—Newm.
  Cidaria populata.—Newm., Sta.
  Carsia paludata (imbutata).—Ent. Mag. ix. 93.

Vaccinium Myrtillus : WHORTLEBERRY; BILBERRY.—100.
  Bombyx rubi.
  Lasiocampa ilicifolia.—Newm., Sta.
  Acronycta auricoma.—Sta.
  Noctua brunnea.—Ent. xxii. 168.
  „   *c-nigrum.*
  „   *baia (baja).*
  „   *sobrina.*
  Hadena rectilinea.—Sta.
  Calocampa solidaginis.—Newm., Sta., Ent. Mag. ix 92.
  Anarta cordigera. —Newm., Sta.
  Epione advenaria.—Newm., Sta.
  Boarmia repandata.—Ent. xvi. 270.
  Halia brunneata (pinetaria).—Newm., Sta., Ent. Mag. v. 108.
  Oporabia filigrammaria.—Ent. ix. 158. In confinement.
  Larentia cæsiata.—Ent. Mag. xii. 6., Ent. ix. 159. In confinement.
  „   *didymata.*—Ent. ix. 159.
  Eupithecia debiliata.—Newm.
  Hypsipetes sordidata (elutata).—Sta.
  Eucosmia undulata.—Ent. xiii. 79. (?).
  Cidaria immanata.—Ent. Mag. xxiii. 279.
  „   *populata.*—Sta.

Vaccinium Oxycoccos : CRANBERRY.—66.
  Carsia paludata (imbutata).—Newm., Sta., Ent. Mag. ix. 93.

Vaccinium uliginosum : BOG WHORTLEBERRY.—17.
  Anarta cordigera.

Arctostaphylos Uva-ursi : RED BEAR-BERRY.—31.
  Anarta melanopa. (?).

Calluna Erica : LING; HEATHER.—110.
  Emydia cribrum.—Newm., Sta.
 * „   *striata (grammica).*—Newm., Sta.

Bombyx rubi.—Ent. xix. 157.
  „   *quercus.*—Ent. xviii. 288.
Acronycta rumicis.—Ent. xviii. 288.
  „   *menyanthidis.*—Sta., Ent. Mag. vii. 88 ; ix. 43.
Agrotis agathina.—Newm.
  „   *strigula (porphyrea).*—Newm.
  „   *ashworthii.*—Sta. In confinement.
Noctua castanea (neglecta).—Ent. xxii. 49., Newm., Sta.
Epunda lutulenta.—Ent. Mag. vi. 236.
Aplecta occulta.—Ent. Mag. xii. 66.
Anarta myrtilli.—Newm., Sta.
Dasydia obfuscaria.—Ent. Mag. viii. 20.
Boarmia repandata.—Ent. xvi. 270.
Acidalia immorata.—Ent. xx. 289.
  „   *contiguaria.*—Ent. Mag. iii. 69. In confinement.
Scodiona belgiaria.—Newm.
Selidosema ericetaria (plumaria).
Aspilates strigillaria.—Newm.
Oporabia filigrammaria.—Ent. ix. 158.
Larentia cæsiata.—Ent. ix. 159., Ent. Mag. xii. 6.,   . 86.
Eupithecia nanata.—Newm.
  „   „   *v. curzoni.*—Ent. xvii. 231.
  „   *minutata.*—Newm.
Hypsipites sordidata (elutata).—Ent. xvii. 257.

Erica Tetralix : CROSS-LEAVED HEATH.—109.
Bombyx rubi.—Newm., Sta.
Saturnia pavonia (carpini).—Newm., Sta.
Bomolocha fontis (crassalis).—Sta.
Boarmia cinctaria.—Newm., Sta.
Acidalia fumata.—Newm.
  „   *contiguaria.*—Ent. xi. 242.
Scodiona belgiaria.—Sta.
Selidosema ericetaria (plumaria).
Pachycnemia hippocastanaria.—Newm., Sta.
Oporabia filigrammaria.—Sta.
Larentia cæsiata.—Sta.
Eupithecia nanata.—Sta.
  „   *minutata.*—Sta.
Eubolia plumbaria (palumbaria).—Sta.

Erica cinerea : COMMON PURPLE HEATH.—108.
Nemeophila russula.—Newm.
Bombyx rubi.—Newm., Sta.
Saturnia pavonia (carpini).—Newm., Sta.
Bomolocha fontis (crassalis).—Sta.
Boarmia cinctaria.—Newm., Sta.
Acidalia fumata.—Newm.
  „   *contiguaria.*—Ent. xi. 242.

Erica cinerea—*continued.*
Scodiona belgiaria.—Sta.
Selidosema ericetaria (plumaria).
Pachycnemia hippocastanaria.—Newm., Sta.
Oporabia filigrammaria.—Sta.
Larentia cæsiata.—Sta.
Eupithecia nanata.—Sta.
„ *minutata.*—Sta.
Eubolia plumbaria (palumbaria).—S:a.

Menziesia cærulea.—SCOTTISH MENZIESIA.—1.
Anarta melanopa.

Armeria maritima: THRIFT.—108.
Sesia musciformis.—Newm., Buckl.
Triphæna comes (orbona).—Ent. xv. 237.
Polia xanthomista.—Newm., Ent. Mag. x. 89.
Epunda nigra.—Ent. xv. 237

Primula acaulis: PRIMROSE.—111.
Nemeobius lucina.—Newm., Sta.
Axylia putris.—(?).
Xylophasia rurea.—Newm., Sta.
Noctua triangulum.
Triphæna ianthina.—Sta.
„ *fimbria.*—Sta.
Aplecta occulta.—Newm.
Melanippe montanata.—Newm., Sta.

Primula veris: COWSLIP.—89.
Nemeobius lucina.—Buckl., Newm., Sta.
Xylophasia rurea.—Newm., Sta.
Tryphæna orbona (subsequa).—Ent. Mag. ix. 57.

Lysimachia vulgaris: LOOSESTRIFE.—78.
Collix sparsata.—Newm., Sta., Ent. xii. 59.

Anagallis arvensis: SCARLET PIMPERNEL.—96.
Acidalia dilutaria (osseata).—Ent. Mag. v. 96. In confinement.

Fraxinus excelsior: ASH.—109.
Sphinx ligustri.—Buckl.
Sciopteron tabaniformis (asiliformis).—Newm. In the roots.
Cossus ligniperda. In the wood.
Zeuzera pyrina (æsculi). In the wood.
Pæcilocampa populi.—Newm.
Acronycta ligustri.—Newm., Sta.
Xanthia circellaris (ferruginea).—Ent. xviii. 301.
Cirrhædia xerampelina.—Newm., Sta.
Xylina semibrunnea.
Catocala fraxini.—Newm., Sta.

Fraxinus excelsior—*continued.*
　Selenia tetralunaria (illustraria).—Sta.
　Eugonia (Ennomos) fuscantaria.—Newm., Sta., Ent. iii. 159.
　　In confinement.
　Eupithecia fraxinata.—Newm.
　　„　　*exiguata.*—Newm.
　Lobophora polycommata.—Newm.
　Cidaria siterata (psittacata). Ent. Mag. xvii. 170.

Ligustrum vulgare : PRIVET.—81.
　Sphinx ligustri.—Buckl., Newm., Sta.
　Sesia andreniformis.—Ent. xx. 102.　On the flowers.
　Acronycta ligustri.—Newm., Sta.
　Pericallia syringaria.—Newm., Sta.
　Eugonia (Ennomos) fuscantaria.—Sta.
　Hemerophila abruptaria.
　Lobophora viretata.—Newm., Sta., Ent. x. 98.

Syringa vulgaris : LILAC.
　Sphinx ligustri.—Buckl., Newm., Sta.
　Cossus ligniperda.—Newm.
　Zeuzera pyrina (æsculi).
　Tæniocampa gothica.—Sta.
　Pericallia syringaria.—Newm., Sta.
　Eugonia (Ennomos) quercinaria (angularia).—Newm.
　Hemerophila abruptaria.—Newm., Sta.　Also on rose.
　Acidalia rusticata.　In confinement.

Vinca major : GREATER PERIWINKLE.
　* *Chærocampa nerii.*—Sta.
　Aplecta occulta.—Ent. Mag. xii. 66.　In confinement.

Gentiana amarella : AUTUMNAL GENTIAN.—81.
　Eupithecia scabiosata (subumbrata).—Newm. On the flowers.
　　„　　*satyrata.*—Newm.　On the flowers.

Gentiana campestris : FIELD GENTIAN.—85.
　Eupithecia scabiosata (subumbrata).—Newm. On the flowers.
　　„　　*satyrata.*—Newm,　On the flowers.

Menyanthes trifoliata : BUCK-BEAN ; MARSH TREFOIL.—106.
　Acronycta menyanthidis.—Ent. Mag. vii. 88.

Cynoglossum officinale : HOUND'S-TONGUE.—75.
　Callimorpha dominula.—Newm.
　　„　　*hera.*
　Agrotis ripæ.—Newm.
　Tæniocampa opima.—Ent. Mag. ix. 21.

FOOD-PLANTS AND LARVÆ. 113

Borago officinalis : BORAGE.
 Deiopeia pulchella.—Ent. xi. 186.   In confinement.
 Agrotis vestigialis (valligera).

Symphytum officinale : COMFREY.—85.
 Arctia caia (caja).—Ent. xiii. 172.

Myosotis palustris : FORGET-ME-NOT.—98.
 Deiopeia pulchella.—Buckl., Ent. xi. 186 & 251.   In confinement.

Myosotis arvensis : FIELD FORGET-ME-NOT.—111
 Deiopeia pulchella.—Newm., Sta.
 Uropteryx sambucaria (sambucata).—Newm.

Lithospermum arvense : CORN GROMWELL ; BASTARD
  ALKARET.—86.
 Epunda lutulenta.—Newm.

Echium vulgare : VIPER'S BUGLOSS.—90.
 Vanessa cardui.—Buckl.
 Dianthœcia irregularis.—Newm.

Calystegia Sepium : GREAT BINDWEED.—93.
 Sphinx convolvuli.—Buckl.

Convolvulus arvensis : FIELD BINDWEED.—92.
 Sphinx convolvuli.—Newm., Sta., Buckl., Ent. Mag. ix. 287.
 Agrophila trabealis (sulphuralis).—Newm., Sta.
 Acontia luctuosa.—Newm., Sta., Ent. Mag. v. 75.
 Acidalia emarginata.—Sta.
 Eupithecia pumilata.—Sta.

Solanum dulcamara : WOODY NIGHTSHADE.—66.
 Acherontia atropos.—Buckl.

Lycium barbarum : TEA-TREE.
 Acherontia atropos.—Newm., Sta.

Atropa Belladonna : DEADLY NIGHTSHADE.—33.
 Acherontia atropos.—Newm.

Hyoscyamus niger : HENBANE.—78.
 Heliothis peltigera.—Newm., Sta.

Verbascum Thapsus : MULLEIN.—90.
 Gortyna ochracea (flavago).—Newm.   In the stems.
 Eremobia ochroleuca.—Ent. Mag. xxi. 159.
 Cucullia verbasci.—Newm.

Verbascum Lychnitis : WHITE MULLEIN.—12.
 Cucullia lychnitis.—Newm.   On the flowers and seeds.

Verbascum nigrum : BLACK MULLEIN.—42.
 Cucullia lychnitis.—Newm.   On the flowers and seeds.
   „ *verbasci.*

Verbascum Blattaria.—MOTH MULLEIN.
  Cucullia scrophulariæ.—Newm., Sta.
Linaria Cymbalaria : IVY-LEAVED TOAD-FLAX.
  Epunda lichenea.—Ent. xxii. 139.
Linaria vulgaris : YELLOW TOAD-FLAX.—98.
  Heliothis dipsacea.—Newm., Sta., Ent. Mag. xi. 257. On the flowers and seeds.
  Eupithecia linariata.—Newm., Sta. On the flowers and seeds.
Scrophularia aquatica : WATER FIG-WORT.—70.
  Hepialus humuli.—Ent. xviii. 263.
  Gortyna ochracea (flavago).—Sta. In the stems.
  Cucullia verbasci.—Newm., Ent. ix. 233.
    „   *scrophulariæ.*—Newm., Sta.
Scrophularia nodosa : KNOTTED FIG-WORT.—106.
  Cucullia scrophulariæ.—Newm., Sta., Ent. ix. 233.
    „   *verbasci.*—Ent. ix. 233.
Digitalis purpurea : FOXGLOVE.—107.
  Melitæa athalia.—Ent. xv. 153 ; xix. 156.
  Gortyna ochracea (flavago).—Newm. In the stem.
  Triphæna comes (orbona).—Ent. xv. 237.
  Polia flavicincta (flavocincta).—Ent. xix. 91.
  Euplexia lucipara.—Sta.
  Eupithecia pulchellata.—Newm. On the flowers.
Veronica serpyllifolia : THYME-LEAVED SPEEDWELL.—111.
  Aspilates gilvaria.—Ent. Mag. viii. 116. In confinement.
Veronica Chamædrys : GERMANDER SPEEDWELL.—111.
  Melitæa athalia.—Newm.
Euphrasia officinalis : EYE-BRIGHT.—111.
  Emmelesia adæquata (blandiata).—Newm., Sta.
Bartsia Odontites : RED BARTSIA.—111.
  Emmelesia unifasciata.—Ent. Mag. vi. 186 ; xi. 140.
Pedicularis palustris : MARSH RED RATTLE.—107.
  Spilosoma urticæ.—Ent. Mag. xxi. 207.
Rhinanthus Crista-galli : YELLOW RATTLE.—112.
  Emmelesia albulata.—Newm., Sta. On the seeds.
Melampyrum pratense : COW-WHEAT.—105.
  Melitæa athalia.—Ent. xv. 153., Ent. Mag. xxiii. 21., Buckl.
  Eupithecia plumbeolata.—Newm., Ent. Mag. iii. 45. On the flowers.
Melampyrum sylvaticum : WOOD COW-WHEAT.—21.
  Melitæa athalia.—Ent. Mag. xxiii. 21.

Mentha hirsuta : HAIRY MINT.
  Spilosoma urticæ.—Newm. Also on other *Menthæ.*
  Epunda lutulenta.—Ent. Mag. vi. 236. Also on other *Menthæ.*
  Acidalia ornata. —Ent. Mag. iii. 44. In confinement on this
     and other *Menthæ.*

Origanum vulgare : MARJORAM.—89.
  Acidalia ornata.—Ent. Mag. iii. 44.
  Eupithecia scabiosata (subumbrata).—Newm. On the flowers.
  „   *satyrata.*—Newm. On the flowers.

Thymus Serpyllum : THYME.—III.
  Lycæna arion.—Newm., Buckl.
  Zygæna pilosellæ (minos) v. nubigena.—Buckl., Ent. Mag. v.
     73.
  Agrotis ashworthii.—Newm , Ent. xxiii. 6.
  Hypenodes costæstrigalis.—Ent. Mag. vi. 216. In confine-
     ment.
  Acidalia ornata.—Sta.
     „   *rubiginata (rubricata).*
     „   *straminata.*
  Aspilates gilvaria.—Ent. Mag. viii. 116.
  Eupithecia constrictata.—Newm. On the flowers.

Nepeta Glechoma : GROUND IVY.—102.
  Acidalia marginepunctata (promutata).
  Coremia ferrugata.—Newm.

Prunella vulgaris : SELF-HEAL.—112.
  Eupithecia scabiosata (subumbrata).—Newm. On the flowers.
  „   *satyrata.*—Newm. On the flowers.

Stachys sylvatica : HEDGE WOUND-WORT.—108.
  Acidalia strigilaria.—Newm., Sta.

Galeopsis Ladanum : RED HEMP NETTLE.—73.
  Emmelesia alchemillata.—Newm., Ent. xii. 128. On the
     seeds.

Galeopsis Tetrahit : HEMP NETTLE.—112.
  Emmelesia alchemillata.—Newm., Ent. xii. 128. On the
     seeds.

Lamium purpureum : PURPLE DEAD-NETTLE.—110.
  Hepialus lupulinus.—Newm., Buckl. On the roots.
  Arctia caia (caja).—Newm.
  Plusia bractea.—Ent. xvi. 134.
  Triphæna ianthina.—Ent. xxii. 151.
     „   *fimbria.*—Ent. xxii. 151.
  Venilia macularia (maculata) —Sta (?).

Lamium album : WHITE DEAD-NETTLE.—100.
   Hepialus lupulinus.—Newm. On the roots.
      „   *humuli.*—Newm., Sta. On the roots.
   Callimorpha hera.
   Arctia caia (caja).—Newm.
   Plusia chrysitis.—Newm.
     „   *iota.*—Newm.
     „   *bractea.*—Ent. xvi. 134.
   Triphæna ianthina.—Ent. xxii. 151.
     „   *fimbria.*—Ent. xxii. 151.
   Coremia quadrifasciaria.—Ent. xiv. 117.

Ballota nigra : BLACK HOREHOUND.—76.
   Hepialus lupulinus.—Newm. On the roots.
     „   *humuli*—Newm. On the roots.

Teucrium Scorodonia : WOOD SAGE.—108.
   Melitæa athalia.—Newm.

Plantago : PLANTAIN.
   Melitæa aurinia (artemis).—Sta.
*   „   *dia.*
   Callimorpha hera.
   Nemeophila russula.—Sta.
     „   *plantaginis.*—Newm., Sta.
   Spilosoma fuliginosa.—Newm., Sta.
     „   *mendica.*—Sta.
   Leucania lithargyria.—Sta.
   Laphygma exigua.—Newm.
   Heliophobus hispidus.—Sta.
   Hydrilla palustris.—Newm., Sta.
   Caradrina alsines.—Sta.
   Agrotis saucia.—Newm., Sta.
     „   *aquilina.*—Newm.
   Noctua ditrapezium. In confinement.
     „   *dahlii.*
   Pachnobia leucographa.—Sta.
   Polia flavicincta (flavocincta).—Ent. xix. 128.
   Hyria muricata (auroraria).—Sta.
   Acidalia marginepunctata (promutata).—Ent. xiv. 116
     „   *immutata.*

Plantago major : GREATER PLANTAIN.—112.
   Melitæa athalia.—Buckl., Newm., Sta.
   Carterocephalus palæmon (paniscus).—Newm., Sta.
   Grammesia trigrammica (trilinea).—Newm., Sta..
   Hadena dissimilis (suasa).—Ent. Mag. iii. 136; iii. 90. In confinement.

Plantago lanceolata : NARROW-LEAVED PLANTAIN.—112.
  Melitæa athalia.—Buckl., Newm., Sta.
  * *Syntomis phegea.*
  Bombyx castrensis.—Sta
  Noctua umbrosa.—Ent. Mag. viii. 140. In confinement.

Plantago maritima : SEA PLANTAIN.—78.
  Melitæa cinxia.—Ent. xiii. 38.
  Agrotis tritici.
  Polia xanthomista.—Ent. x. 20., Ent. Mag. x. 89.

Plantago Coronopus : BUCK'S-HORN PLANTAIN.—96.
  Melitæa cinxia.—Ent. xiii. 38.

Chenopodium Vulvaria : STINKING GOOSE-FOOT ; ORACH. —37.
  Mamestra albicolon.—Newm.
  ,, *brassicæ.*—Newm.
  Caradrina morpheus.—Ent. xiii. 93.
  Hadena trifolii (chenopodii).— Newm., Sta.
  ,, *atriplicis.*--Newm., Sta.
  * ,, *peregrina.*—Newm.
  Eupithecia subnotata.—Newm., Sta. On the flowers and seeds.
  Pelurga comitata.—Newm. Sta.

Chenopodium album : WHITE GOOSE-FOOT.—109.
  Mamestra albicolon —Newm.
  ,, *brassicæ.*—Newm.
  Caradrina morpheus.—Ent. xiii. 93.
  Agrotis corticea.—Ent. Mag. viii. 89. In confinement.
  * *Hadena peregrina.*—Newm.
  ,, *trifolii (chenopodii).*—Newm., Sta.
  ,, *atriplicis.*—Newm., Sta.
  Acidalia straminata. Also on other *Chenopodia.*
  Eupithecia subnotata.—Newm., Sta. On the seeds and flowers.
  Pelurga comitata.—Newm., Sta.

Chenopodium Bonus-Henricus : GOOD KING HENRY.—97.
  Mamestra albicolon.—Newm.
  ,, *brassicæ.*—Newm.
  * *Hadena peregrina.*—Newm.
  ,, *trifolii (chenopodii)*.—Newm., Sta.
  ,, *atriplicis* —Newm., Sta.
  Eupithecia subnotata.—Newm., Sta. On the seeds and flowers.
  Pelurga comitata.—Newm., Sta.

Beta maritima : BEET.—35.
  Eupithecia subnotata.—Ent. xvi. 247.

Atriplex patula : ORACH.
  Mamestra albicolon.— Newm.
  Hadena atriplicis.—Newm.
  „ *trifolii (chenopodii).*—Sta.
  Eupithecia subnotata.—Newm.

Atriplex laciniata : SEA ORACH.—39.
  Mamestra albicolon.—Newm.
  Hadena trifolii (chenopodii).—Sta.
  „ *atriplicis.*—Newm.
  Eupithecia subnotata.—Newm.

Salsola Kali : PRICKLY GLASS-WORT.—58.
  * *Hadena peregrina.*—Newm.

Polygonum aviculare : KNOT-GRASS.—110.
  Deilephila livornica.—Buckl. In confinement.
  Lithosia griseola.—Buckl. In confinement.
  Bombyx rubi.—Ent. xxi 275.
  Acronycta rumicis.—Newm.
  Agrotis corticea.—Ent. Mag. viii. 89. In confinement.
    „ *puta.*
    „ *cinerea.*—(?). On the roots.
    „ *simulans (pyrophila).*—(?).
    „ *ashworthii.*—Ent. Mag. xvii. 135.
    „ *lunigera.*—Ent. Mag. iii. 188. In confinement.
  Noctua ditrapezium.—(?).
    „ *dahlii.*—(?).
    „ *festiva.*—Ent. Mag. xi. 139. In confinement
  Tryphæna pronuba.—Ent. Mag. xvii. 135.
  Cerastis erythrocephala.—(?).
  Dasycampa rubiginea.—Ent. Mag. v. 206. In confinement.
  Aplecta prasina (herbida).—Ent. xv. 42.
    „ *occulta.*—Ent. Mag. xi. 157 ; xii. 66. In confine-
        ment.
    „ *advena.*—Newm.
  Hadena trifolii (chenopodii).—Ent. xv. 273.
    „ *atriplicis.*—Newm. Also on other *Polygona.*
    „ *dissimilis (suasa).*—Newm., Ent. Mag. iii. 90 ; iii.
        136.
    „ *thalassina.*—Newm.
    „ *genistæ.*—Newm. Also on other *Polygona.*
  Xylomiges conspicillaris.—Ent. Mag. xii. 83.
  Heliothis umbra (marginatus).—Newm. In confinemeut.
  * *Acontia solaris.*—(?).
  * *Plusia ni.*—(?).
  Dasydia obfuscaria.—Ent. Mag. viii. 20. In confinement.
  Hyria muricata (auroraria).—Ent. ix. 198 ; xv. 41.
  Timandra amataria.—Newm., Sta. Also on other *Polygona.*

Polygonum aviculare—*continued.*
  Lythria purpuraria.—Newm.   Also on other *Polygona.*
  Sterrha sacraria.—Newm.   In confinement.
  Acidalia dilutaria (osseata).—Ent. xi. 19.   In confinement.
  ,,   *virgularia.*—Ent. xi. 91., Ent. Mag. ix. 246. In confinement.
  ,,   *straminata.*—Ent. Mag. xi. 116.   In confinement.
  ,,   *degenaria.*—Ent. Mag. ix. 115.   In confinement.
  ,,   *strigilata.*—Ent. Mag. viii. 91 ; ix. 197.   In confinement.
  ,,   *trigeminata.*—Ent. Mag. viii. 22 ; ix. 197.   In confinement.
  ,,   *emutaria.*—Ent. Mag. ix. 197.   In confinement.
  ,,   *holosericata.*—Ent. Mag. ix. 197.   In confinement.
  ,,   *remutaria.*   In confinement.
  ,,   *circellata.*—Ent. Mag. iii. 90.   In confinement.
  ,,   *subsericeata.*—Ent. Mag. iii. 90.   In confinement.
  ,,   *contiguaria.*—Ent. Mag. iii. 69.   In confinement.
  ,,   *bisetata.*—Ent. Mag. v. 96.   In confinement.
  ,,   *inornata.*   In confinement.
  Scoria lineata (dealbata).—Newm.

Polygonum Persicaria : COMMON PERSICARIA.—112.
  Camptogramma fluviata.—Newm.

Rumex : DOCK.
  Deilephila livornica.—Buckl.   In confinement.
  Sesia chrysidiformis.—Ent. xvii. 23 ; xx. 105.   On the roots.
  Hepialus sylvanus.—Buckl., Ent. Mag. iii. 136.   On the roots.
  ,,   *humuli.*—Buckl.   On the roots.
  Spilosoma mendica.—Newm., Sta.
  ,,   *lubricepeda.*—Newm.
  ,,   *menthastri.*
  Caradrina morpheus.—Newm.
  ,,   *alsines.*—Sta.
  Xylophasia rurea.—Newm., Sta.
  Dipterygia scabriuscula (pinastri).—Newm., Sta.
  Mamestra brassicæ.—Newm.
  Agrotis saucia.—Newm., Sta.
  ,,   *cinerea.*   In confinement.
  ,,   *corticea.*—Ent. Mag. viii. 89.   In confinement.
  ,,   *obscura (ravida).*—Sta.
  Noctua glareosa.—Newm.
  ,,   *umbrosa.*—Ent. Mag. viii. 140.   In confinement.
  ,,   *triangulum.*   In confinement.
  ,,   *dahlii.*   In confinement.
  ,,   *ditrapezium.*   In confinement.
  ,,   *c-nigrum.*   In confinement.

Rumex—*continued.*
   Triphæna interjecta.—Newm.
   Mania typica.—Newm., Sta.
    ,, *maura.*—Sta.
   Tæniocampa gothica.—Ent. xix. 273.
    ,, *incerta (instabilis).*—Newm.
   Pachnobia rubricosa.—Newm., Sta.
   Anchocelis pistacina.—Sta.
   Polia flavicincta (flavocincta).—Ent. xix. 276.
   Epunda nigra.—Sta.
   Aplecta prasina (herbida) —Newm.
    ,, *nebulosa.*—Sta.
   Hadena atriplicis.—Newm., Sta.
    ,, *oleracea.*—Newm.
   Calocampa vetusta.—Newm.
   Timandra amataria.—Newm., Sta.
   Sterrha sacraria.—Newm.
Rumex crispus: CURLED DOCK.—109.
   Polia chi. In confinement.

Rumex pulcher: FIDDLE DOCK.—42.
   Polyommatus phlæas.—Newm.
   Aplecta occulta.—Ent. Mag. xii. 66. In confinement.

Rumex obtusifolius: BROAD-LEAVED DOCK.—109.
   Polyommatus phlæas.—Newm.
*   ,, *virgaureæ.*—Sta.

Rumex aquaticus :—37.
   Polyommatus dispar.—Sta.

Rumex Hydrolapathum: WATER DOCK.—68.
   Polyommatus dispar.—Newm., Sta.

Rumex Acetosa: SORREL.—112.
   Polyommatus phlæas.—Newm., Sta., Buckl.
*   ,, *virgaureæ.*—Sta.
   Sesia chrysidiformis.—Ent. xvii. 23 ; xx. 105., Buckl. On the roots.
   Ino statices.—Newm., Sta., Buckl.
    ,, *geryon.*—(?).
   Zygæna exulans.—Buckl. In confinement.
* *Syntomis phegea.*
   Leucania comma.—Sta.
   Agrotis cinerea.
   Noctua glareosa.—Newm.
    ,, *depuncta.*—Newm., Sta.
   Timandra amataria.—Newm.

Rumex Acetosella: SHEEP SORREL.—112.
   Polyommatus phlœas.—Newm., Buckl.
   Sesia chrysidiformis.—Ent. xvii. 23 ; xx. 105., Buckl. On the roots.
   Ino geryon.—Newm. In confinement.
   „ *statices.*—Newm.
   Leucania comma.—Sta.
   Noctua glareosa.—Newm.
   „ *depuncta.*—Newm., Sta.
   Erastria venustula.—(?). Ent. xvi. 138.
   Acidalia imitaria.—Newm. In confinement.

Euphorbia Esula.
   Agrotis cursoria.—Newm., Sta., Ent. Mag. ix. 14.

Euphorbia Cyparissias: CYPRESS SPURGE.
   Deilephila euphorbiæ.—Buckl., Sta., Ent. Mag. xi. 73.
   Minoa murinata (euphorbiata).—Newm.

Euphorbia Paralias: SEA SPURGE.—29.
   Deilephila euphorbiæ.—Buckl., Newm., Sta., Ent. Mag. xi. 73.
   Minoa murinata (euphorbiata).—Sta.

Euphorbia portlandica: PORTLAND SPURGE.—19.
   Deilephila euphorbiæ.—Buckl., Ent. Mag. xi. 73.

Euphorbia Peplus: PETTY SPURGE.—98.
   Deilephila euphorbiæ.—Buckl., Ent. Mag. xi. 73. In confinement.
   Minoa murinata (euphorbiata).—Sta.

Mercurialis perennis: DOG'S MERCURY.—105.
   Plusia bractea.—Ent. xiv. 222.

Ulmus montana: WYCH ELM.—93.
   Vanessa polychloros.—Newm.
   Thecla w-album.—Newm., Ent. ix. 158.
   Acronycta alni.—Ent. x. 74 ; xii. 251.
   Xanthia gilvago.—Ent. ix. 158 ; xviii. 319.
   „ *circellaris (ferruginea).*—Ent. xviii. 319.
   Asthena blomeri (blomeraria).—Ent. Mag. xi. 87.
   Abraxas sylvata (ulmata).

Ulmus campestris: ELM.
   Vanessa c-album.—Newm., Sta.
   „ *polychloros.*—Newm., Sta.
   Thecla w-album.—Newm., Sta.
   Smerinthus tiliæ.—Newm., Sta., Buckl.
   Sesia asiliformis (cynipiformis).—Newm. In the bark.
   Cossus ligniperda.—Buckl., Newm. In the wood.
   Zeuzera pyrina (æsculi).—Newm., Sta. In the wood.
   Eriogaster lanestris.—Newm.

Ulmus campestris—*continued.*
   Phalera bucephala.—Newm.
   Tæniocampa stabilis —Sta.
      „   *munda.*—Sta.
   Cerastis vaccinii.—Newm.
   Scopelosoma satellitia.—Sta.
   Xanthia gilvago.—Newm., Sta., Ent. Mag. iii. 45. On the seeds.
   Calymnia diffinis.—Newm., Sta.
      „   *affinis.*—Newm., Sta.
   Miselia bimaculosa.—Sta.
   Amphipyra pyramidea.—Sta.
   Asteroscopus nubeculosa.—(?)., Sta.
   Metrocampa margaritaria.—Newm.
   Selenia lunaria.—Sta.
   Eugonia (Ennomos) quercinaria (angularia).—Newm.
   Biston hirtaria.—Sta.
   Amphidasys betularia.—Ent. xix. 254.
   Tephrosia crepuscularia.—Sta.
   Abraxas sylvata (ulmata).—Newm., Sta.
   Hybernia defoliaria.—Ent. xxi. 212.
   Anisopteryx œscularia.—Newm.

Humulus Lupulus : Hop.—82.
   Vanessa c-album.—Newm., Sta., Buckl.
   Hepialus humuli.—Sta. On the roots.
   Dasychira pudibunda.—Newm., Sta.
   Caradrina morpheus.—Ent. xiii. 93.
   Axyla putris.—Ent. xiv. 44.
   Hadena porphyrea (satura).—Ent. Mag. xxii. 63.
   Habrostola triplasia.—Newm.
   Plusia gamma.—Newm.
   Hypena rostralis.—Sta.
   Eupithecia assimilata.—Ent. xviii. 142.

Urtica dioica : Stinging Nettle.—112.
   Vanessa c-album.—Buckl., Sta.
      „   *urticæ.*—Newm., Sta.
      „   *antiopa* —Newm.
      „   *io.*—Newm., Sta.
      „   *atalanta.*—Newm., Sta.
   Spilosoma mendica.—Ent. Mag. xxiii. 187.
      „   *menthastri.*—Ent. xxi. 215.
   Axylia putris.—Ent. xiv. 44.
   Hadena oleracea.—Newm.
   Habrostola tripartita (urticæ).—Newm., Sta.
      „   *triplasia.*—Newm., Sta.

Urtica dioica—*continued.*
   Plusia chrysitis.—Newm., Sta.
     ,,   *iota.*—Newm., Sta.
     ,,   *pulchrina.*—Sta.
     ,,   *interrogationis.*—Newm., Sta.
   Hypena proboscidalis.—Sta.
   Venilia macularia (maculata).—Sta.

Urtica urens: SMALL NETTLE.—107.
   Vanessa urticæ.—Newm.
   Plusia pulchrina.—Sta.

Parietaria officinalis: PELLITORY-OF-THE-WALL.—91.
   Vanessa atalanta.—Buckl.
   Acidalia rusticata.

Myrica Gale: SWEET GALE.—79.
   Acronycta menyanthidis.—Newm., Sta.
     ,,   *euphorbiæ.*—Sta., Ent. Mag. vii. 83
   Noctua subrosea.—Newm., Sta.
   Hadena contigua.—Ent. xvii. 171.
   Xylina lambda.
   Melanippe hastata.—Newm.

Betula alba: BIRCH.—107.
   Vanessa antiopa.—Buckl., Newm.
   Thecla rubi.—Ent. Mag. x. 212.
   Sesia scoliiformis.—Buckl., Newm., Sta.   In the bark.
     ,,   *culiciformis.*—Buckl., Newm., Sta.   In the bark.
   Hylophila prasinana.—Sta.
   Zeuzera pyrina (æsculi).—Buckl.   In the wood.
   Heterogenea limacodes (testudo).—Ent. xix. 133.
     ,,   *asella (asellus).*—Ent. Mag. xvii. 169.
   Psilura monacha.—Newm.
   Dasychira pudibunda.—Ent. xxi. 4.
   Orgyia gonostigma.—Ent. xii. 106.
   Trichiura cratægi.
   Pœcilocampa populi.—Buckl.
   Spilosoma mendica.—Ent. xviii. 194.
   Endromis versicolor—Buckl., Newm., Sta.
   Drepana (Plaptypteryx) lacertinaria (lacertula).—Newm., Sta.
       ,,   ,,   *harpagula (sicula)* —Sta.
       ,,   ,,   *falcataria (falcula).*—Newm., Sta.
       ,,   ,,   *binaria (hamula).*—Newm., Sta.
   Dicranura bicuspis.—Buckl., Sta.
   Stauropus fagi.—Newm., Sta., Ent. ix. 270.
   Lophopteryx camelina.—Newm., Sta.
     ,,   *carmelita.*—Newm., Sta.

Betula alba—*continued.*
  Notodonta bicolor.—Buckl.
      „   *dictæoides.*—Buckl., Newm., Sta.
      „   *dromedarius.*—Buckl., Newm., Sta.
      „   *trilophus.*—Sta.
      „   *trimacula (dodonea).*—Sta.
  Cymatophora duplaris.—Newm., Sta.
      „   *fluctuosa.*—Newm., Sta.
  Asphalia diluta.—Newm.
      „   *flavicornis.*—Newm., Sta.
  Moma (Diphthera) orion.—Newm.
  Demas coryli.—Sta.
  Acronycta leporina.—Newm., Sta., Ent. Mag. xi. 158.
      „   *alni.*—Sta.
      „   *auricoma.*—Sta.
      „   *euphorbiæ.*—Ent. Mag. vii. 83. In confinement.
  Noctua ditrapezium.—Ent. xi. 141. In confinement.
      „   *triangulum.*
  Cosmia palacea (fulvago).—Newm., Sta.
  Calymnia trapezina.—Sta.
  Aplecta nebulosa.—Newm.
      „   *tincta.*—Newm., Sta.
  Hadena contigua.—Newm.
  Xylina furcifera (conformis).—Newm.
      „   *lambda.*
  Asteroscopus nubeculosa.—Buckl., Newm., Sta.
  Pechypogon barbalis.—Sta.
  Brephos parthenias.—Newm., Sta.
  Metrocampa margaritaria.—Newm., Sta.
  Eurymene dolobraria.—Sta.
  Selenia tetralunaria (illustraria).—Newm., Sta.
  Eugonia (Ennomos) autumnaria (alniaria).—Newm., Sta., Ent. ix. 278, Ent. Mag. iii. 159.
      „   „ *alniaria (tiliaria).*—Newm., Sta., Ent. Mag. iii. 159.
      „   „ *erosaria.*—Newm., Sta.
      „   „ *quercinaria (angularia).*—Newm., Ent. Mag. iii. 159.
  Amphidasys strataria (prodomaria).—Newm., Sta.
      „   *betularia.*—Newm., Sta.
  Boarmia repandata.—Newm., Sta.
      „   *gemmaria (rhomboidaria).*—Newm.
      „   *abietaria.*—Ent. xxii. 287.
  Tephrosia consonaria.—Newm., Ent. Mag. ix. 17.
      „   *crepuscularia.*—Ent. xix. 271.
      „   *biundularia.*—Ent. xix. 159.
      „   *luridata (extersaria).*—Newm., Sta.
      „   *punctularia.*—Newm., Sta.

FOOD-PLANTS AND LARVÆ. 125

Betula alba—*continued.*
Cleora angularia (viduaria)—(?).
Geometra papilionaria.—Newm., Sta.
Iodis lactearia.—Sta.
Zonosoma (Ephyra) pendularia.—Newm., Sta., Ent. Mag. x. 71.
Asthena sylvata.—Ent. xiii. 14.
Cabera pusaria.—Newm., Sta.
„ *rotundaria.*—Sta.
Fidonia carbonaria.—Newm.
Hybernia aurantiaria.—Newm., Sta., Ent. Mag. viii. 90.
„ *marginaria (progemmaria).*—Sta.
Cheimatobia brumata.
„ *boreata.*—Newm., Sta.
Oporabia autumnaria.—Sta.
Melanippe hastata.—Newm., Sta.
Cidaria miata.—Newm., Sta.
„ *truncata (russata).*—Newm.
„ *testata.*—Newm.

Alnus glutinosa : ALDER.—109.
Sesia sphegiformis.—Buckl., Newm., Sta., Ent. Mag. x. 161.
In the stems and suckers.
„ *culiciformis.*—Sta. In the bark and wood.
Hylophila prasinana.—Sta.
Zeuzera pyrina (æsculi).
Drepana (Platypteryx) falcataria (falcula).—Sta.
Dicranura bicuspis.—Buckl., Newm.
Lophopteryx camelina.—Buckl.
Notodonta dromedarius.—Ent. xvii. 35., Buckl.
Acronycta alni.—Newm., Sta.
„ *leporina.*—Ent. xix. 133., xxii. 115.
Cymatophora duplaris.—Ent. Mag. xvii. 171.
Xylina lambda.
„ *furcifera (conformis).*—Newm., Ent. Mag. viii. 114
Epione apiciaria.—Sta.
Eugonia (Ennomos) autumnaria (alniaria).—Sta.
Amphidasys betularia.—Ent. xix. 254.
Tephrosia crepuscularia.—Sta.
„ *punctularia.*—Sta.
Zonosoma orbicularia.—Sta.
Asthena sylvata.—Sta., Ent. xii. 296.
„ *luteata.*—Ent. Mag. xxiii. 109 ; xxiii. 141.
Eupisteria obliterata (heperata).
Cabera pusaria.—Ent. xix. 43.
„ *rotundaria.*—(?).
„ *exanthemaria.*—Newm.
Macaria alternata.—Ent. Mag. xvii. 170.

Alnus glutinosa—*continued.*
   Eupithecia exiguata.—Newm.
   Hypsipetes trifasciata (impluviata).—Newm., Sta., Ent. Mag.
      vii. 42 ; ix. 248.
      ,,   *sordidata (elutata).*—Sta.
   Melanthia bicolorata (rubiginata).—Sta.
   Cidaria miata.—Newm., Sta.

Carpinus betulus: HORNBEAM.—35.
   Calymnia trapezina.—Newm.
   Metrocampa margaritaria—Newm., Sta.
   Tephrosia consonaria.—Newm.
   Asthena candidata—Sta.
   Hybernia marginaria (progemmaria).—Newm.
      ,,   *defoliaria.*—Newm.
      ,,   *aurantiaria.*
   Cheimatobia brumata.—Newm.
   Oporabia dilutata.—Newm.

Corylus avellana: HAZEL.
   Smerinthus tiliæ.—Ent. xxi. 232.
   Hylophila prasinana.—Sta.
   Dasychira fascelina—Newm.
      ,,   *pudibunda.*—Newm.
   Orgyia gonostigma.—Newm., Sta.
      ,,   *antiqua.*
   Bombyx neustria v. bilineatus.
   Lophopteryx camelina.—Sta.
   Phalera bucephala.—Newm., Sta.
   Demas coryli.—Newm.
   Acronycta alni.—Ent. ix. 232 ; x. 141 ; xv. 291.
   Hadena contigua.—Sta.
   Epione paralellaria (vespertaria).—Newm., Sta.
      ,,   *apiciaria.*—Newm.
   Geometra papilionaria.—Sta.
   Asthena candidata.
   Cabera pusaria.—Newm.
   Hybernia defoliaria.—Newm.
   Cheimatobia brumata.—Newm.

Quercus Robur: OAK.—105.
   Thecla quercus.—Newm., Sta.
   Sesia asiliformis (cynipiformis).—Buckl., Newm , Sta. In the bark.
   Hylophila prasinana.—Sta.
      ,,   *bicolorana (quercana).*—Sta.
   Lithosia lurideola (complanula).—Buckl. " Said to feed on lichen."
   Cossus ligniperda.— Buckl., Newm., Sta. In the wood.

Quercus Robur—*continued.*
   Heterogenea asella (asellus).—Newm., Sta.
      „   *limacodes (testudo).*—Newm., Sta.
   Nola confusalis (cristulalis).—Newm., Sta.
     „ *strigula.*—Buckl., Newm., Sta , Ent. Mag. ix. 15.
   Porthesia similis (auriflua).—Sta.
* *Laria l-nigrum (v-nigrum).*—Sta.
   Psilura monacha.—Newm., Sta.
   Dasychira pudibunda.—Newm.
   Orgyia gonostigma.—Newm., Sta.
     „ *antiqua.*
   Trichiura cratægi.—Ent. xv. 234., Ent. Mag. xviii. 38.
   Pœcilocampa populi.—Sta.
   Drepana (Platypteryx) harpagula (sicula).—Sta.
     „   „ *falcataria (falcula).*—Sta.
     „   „ *binaria (hamula).*—Buckl., Newm., Sta.
   Stauropus fagi.—Buckl., Newm., Sta., Ent. ix. 270.
   Lophopteryx camelina.—Buckl., Newm.
   Notodonta trepida.—Newm., Sta.
     „ *chaonia.*—Buckl., Newm., Sta.
     „ *trimacula (dodonæa).*—Buckl., Newm., Sta.
   Phalera bucephala.—Sta.
   Asphalia diluta.—Newm., Sta.
     „ *ridens.*—Newm., Sta.
   Moma orion.—Newm., Sta.
   Demas coryli.—Sta.
   Acronycta aceris.—Newm.
     „ *alni.*—Sta., Ent. ix. 204; xi. 141.
     „ *auricoma.*—Ent. Mag. iii. 261.
   Tæniocampa gothica.—Newm.
     „ *incerta (instabilis)* —Newm., Sta.
     „ *stabilis.*—Newm., Sta.
     „ *miniosa.*—Newm., Sta.
     „ *munda.*—Newm.
     „ *pulverulenta (cruda).*—Newm., Sta.
   Anchocelis rufina.—Newm.. Sta.
   Cerastis vaccinii.—Newm., Sta.
   Scopelosoma satellitia.—Newm., Sta.
   Dasycampa rubiginea.—Sta.
   Oporina croceago.—Newm., Sta.
   Cosmia palacea (fulvago).—Newm., Sta.
   Dicycla oo.—Newm., Sta.
   Calymnia trapezina.—Newm., Sta.
   Hadena protea.—Sta.
     „ *contigua.*—Newm.
   Xylina ornithopus (rhizolitha.)—Sta., Ent. Mag. xii. 140.
     „ *socia (petrificata).*—Newm., Sta.
     „ *semibrunnea.*

Quercus Robur—*continued.*
Agriopis aprilina.—Newm., Sta.
Asteroscopus nubeculosa.—Buckl.
    „   *sphinx (cassinea).*—Newm., Sta.
Amphipyra pyramidea.—Newm., Sta.
Catephia alchymista.—Newm.
Catocala sponsa.—Newm., Sta.
    „   *promissa.*—Sta.
Ophiodes lunaris.—Newm., Sta.
Zanclognatha emortualis. Dead leaves preferred.
Pechypogon barbalis.—Sta.
Brephos parthenias.—Sta.
Herminia derivalis. On the dead leaves in confinement.
Uropteryx sambucaria (sambucata).—Sta.
Metrocampa margaritaria.—Newm., Sta.
Eurymene dolobraria.—Newm., Sta., Ent. Mag. xi. 158.
Selenia lunaria.—Sta.
    „   *tetralunaria (illustraria).*—Ent. xxii. 73., Newm., Sta.
Odontopera bidentata.—Sta
Eugonia (Ennomos) alniaria (tiliaria).—Newm., Sta.
    „   „   *autumnaria (alniaria).*—Ent. ix. 278.
    „   „   *erosaria.*—Newm., Sta., Ent. Mag. iii. 159.
    „   „   *quercinaria (angularia).*—Newm., Sta.
Himera pennaria.—Newm., Sta.
Phigalia pedaria (pilosaria).—Newm., Sta.
Nyssia hispidaria.—Newm.
Amphidasys strataria (prodomaria).—Newm., Sta.
    „   *betularia.*—Newm.
Boarmia gemmaria (rhomboidaria).—Sta.
    „   *roboraria.*—Newm., Sta.
    „   *consortaria.*—Newm., Sta.
Tephrosia crepuscularia.—Ent. xix. 268 ; Ent. Mag. viii. 209.
    „   *biundularia.*—Newm.
    „   *consonaria.*—Ent. Mag. ix. 17.
Phorodesma pustulata (bajularia).—Sta.
Hemithea strigata (thymiaria).—Sta.
Zonosoma (Ephyra) porata.—Sta.
    „   *(Ephyra) punctaria.*—Newm., Sta., Ent. Mag. viii. 183.
Cabera pusaria.—Newm.
Hybernia rupicapraria.—Newm.
    „   *leucophæaria.*—Newm., Sta.
    „   *aurantiaria.*—Newm., Sta.
    „   *marginaria (progemmaria).*—Sta.
    „   *defoliaria.*—Newm.

Quercus Robur—*continued.*
   Anisopteryx æscularia.—Newm.
   Cheimatobia brumata.—Ent. xxi. 188.
   Oporabia dilutata.—Newm., Sta.
   Eupithecia irriguata.—Ent. xviii. 109., Ent. Mag. vii. 15 ;
      viii. 69.
     ,, *abbreviata.*—Newm., Sta.
     ,, *dodoneata.*—Newm., Sta.
   Cidaria siterata (psittacata).—Newm., Ent. Mag. xi. 158.
     ,, *miata.*—Newm., Sta.

Castanea sativa: CHESTNUT.
   Dasychira pudibunda.—Newm.
   Saturnia pavonia (carpini).—Ent. ix. 160. In confinement.
   Acronycta alni.—Ent. xii. 270 ; xv. 235., Ent. Mag. xix. 90.
   Herminia cribralis. On the dead leaves in confinement.

Fagus sylvatica: BEECH.—64.
   Hylophila prasinana.—Sta.
   Nola confusalis (cristulalis).—Ent. Mag. xvii. 169.
   Heterogenea limacodes (testudo).—Sta.
     ,, *asella (asellus).*—Buckl., Sta., Ent. Mag. x. 70.
  * *Laria l-nigrum (v-nigrum).*—Sta.
   Drepana (Platypteryx) cultraria (unguicula).—Newm., Sta.
   Dicranura bicuspis.—Sta.
     ,, *furcula.*—Ent. xxi. 275.
   Stauropus fagi.—Buckl., Sta., Ent. ix. 270.
   Acronycta alni.—Ent. Mag. xvii. 171.
   Orthosia macilenta.—Newm., Sta.
   Scopelosoma satellitia.—Sta.
   Xanthia aurago.—Newm., Sta.
   Brephos parthenias.—Sta.
   Angerona prunaria.—Newm.
   Metrocampa margaritaria—Sta.
   Eurymene dolobraria.—Newm.
   Selenia tretralunaria (illustraria).—Sta.
   Crocallis elinguaria.—Newm.
   Eugonia (Ennomos) autumnaria (alniaria).—Newm.
   Amphidasys betularia.—Ent. xix. 254.
   Tephrosia consonaria.—Newm., Sta.
   Geometra papilionaria.—Sta.
   Zonosoma linearia (trilinearia).—Sta.
   Eupithecia irriguata.—Ent. Mag. vii. 15.

Salix alba: WHITE WILLOW.—86.
   Vanessa antiopa.—Newm., Sta.
     ,, *polychloros.*—Sta., Ent. xxi. 255.
   Smerinthus ocellatus.—Buckl., Newm., Sta.
   Earias chlorana.—Sta. In the shoots.

Salix alba—*continued.*
Cossus ligniperda.—Newm., Sta. In the wood.
Leucoma salicis.—Sta.
Lasiocampa quercifolia.—Newm., Sta.
Saturnia pavonia (carpini).—Newm.
Drepana (Platypteryx) falcataria (falcula).—Sta.
Dicranura vinula.—Newm., Sta., Buckl.
„ furcula.—Ent. xviii. 183.
Notodonta dictæa.—Sta.
„ ziczac.—Buckl.
Demas coryli.—Sta.
Acronycta leporina.—Ent. xix. 133.
„ menyanthidis.—Ent. Mag. ix. 43.
„ alni.—Sta.
Noctua subrosea.—Ent. Mag. xi. 67. In confinement.
Tæniocampa incerta (instabilis).—Sta.
„ gracilis.—Sta.
Amphipyra pyramidea.—Sta.
Orthosia upsilon.—Newm., Sta.
Anchocelis litura.—Sta.
Cleoceris viminalis.—(?).
Xylina semibrunnea.
Gonoptera libatrix.—Sta., Ent. Mag. vii. 117.
Anarta myrtilli.—Ent. Mag. vii. 88. In confinement.
Catocala nupta.—Sta.
Madopa salicalis.— Sta.
Epione apiciaria.—Newm., Sta.
Selenia bilunaria (illunaria).—Newm., Sta.
Amphidasys betularia.— Ent. xix. 254.
Tephrosia crepuscularia.—Sta.
Acidalia inornata.—Newm. On the low shoots.
Scodiona belgiaria.—Ent. xvii. 5.
Larentia cæsiata.—Ent. Mag. vii. 88. In confinement.
Eupithecia vulgata.—Sta.
Lobophora sexalisata (sexalata).—Sta.
Hypsipetes ruberata.—Sta.
Cidaria populata.—Ent. Mag. vii. 88. In confinement.

Salix fragilis : CRACK WILLOW.—84.
Noctua subrosea.—Ent. Mag. xi. 67. In confinement.
Orthosia upsilon.—Newm.
„ lota.—Newm.
Catocala nupta.—Newm.

Salix triandra : THREE-STAMENED WILLOW.—63.
Sesia formiciformis.—Buckl., Sta. In the wood.
Trochilium crabroniformis (bembeciformis).—(?). In the wood.
Earias chlorana.—(?). In the shoots.

Salix viminalis : OSIER.—80.
   Vanessa polychloros.—Newm.
   Sesia formiciformis.—Buckl., Newm., Sta. In the wood.
   Trochilium crabroniformis (bembeciformis).—Newm. In the wood.
   Earias chlorana. In the shoots.
   Dicranura furcula.— Buckl.
   Polia chi.—Ent. Mag. ix. 290. In confinement.

Salix cinerea : SALLOW.—102.
   Thecla quercus.—Ent. x. 285.—(?).
   Smerinthus ocellatus—Buckl., Ent xiii. 172.
     ,,   *populi.*—Sta.
   Trochilium crabroniformis (bembeciformis). Sta., Ent. Mag. x. 161. In the wood.
   Sarothripus undulanus (revayana).—Sta.
   Ocneria dispar.—Ent. xviii. 243.
   Dasychira pudibunda.—Ent. xxii. 151.
   Orgyia antiqua.—Buckl. In confinement.
     ,,   *gonostigma.*—Ent. xii. 106.
   Trichiura cratægi.—Sta.
   Lasiocampa ilicifolia.—Sta.
   Dicranura furcula.—Buckl., Newm.
     ,,   *vinula.*—Buckl., Sta.
   Pterostoma (Ptilodontis) palpina.—Buckl., Newm., Sta.
   Notodonta ziczac.—Buckl., Newm., Sta.
   Pygæra curtula.—Sta.
     ,,   *pigra (reclusa).*—Newm. Sta.
   Acronycta leporina.—Ent. xix. 133 ; xxii. 114.
     ,,   *euphorbiæ.*—Ent. Mag. vii. 83. In confinement.
     ,,   *alni.*—Sta.
   Caradrina morpheus.—Ent. Mag. x. 254.
   Tryphæna comes (orbona).—Newm. After hybernation.
   Noctua brunnea.—Newm.
     ,,   *triangulum.* In confinement.
     ,,   *festiva.*—Newm.
     ,,   *ditrapezium.* In confinement.
     ,,   *castanea v. neglecta.*—Ent. Mag. xxii. 242.
   Xanthia fulvago (cerago).—Ent. Mag. vi. 262.
     ,,   *flavago (silago).*—Ent. Mag. vi. 262.
   Cleoris viminalis.—Newm.
   Polia chi.—Ent. Mag. ix. 290. In confinement.
   Aplecta occulta.—Ent. Mag. xii. 66. In confinement.
   Asteroscopus sphinx (cassinea).—Sta.
   Madopa salicalis.—Sta.
   Brephos notha.—Sta.

Salix cinerea—*continued.*
Eugonia (Ennomos) autumnaria (alniaria).—Ent. Mag. iii.
159. In confinement
„ „ *alniaria (tiliaria).*—Ent. Mag. iii. 159. In confinement.
Zonosoma (Ephyra) orbicularia.—Sta., Ent. x. 98.
Nyssia zonaria.—Ent. Mag. xxiii. 61.
Acidalia rubiginata (rubricata).
Cabera exanthemaria.—Newm., Sta.
Macaria notata.—Newm.. Sta.
Numeria pulveraria.—Newm., Sta.
Fidonia carbonaria.—Newm.
Lomaspilis marginata.—Newm., Sta.
Oporabia filigrammaria.—Newm.
Eupithecia tenuiata.—Newm., Sta. In the catkins.
„ *exiguata.*—Newm.
Lobophora halterata (hexapterata).—Newm., Sta.
Hypsipetes ruberata.—Sta.
„ *sordidata (elutata).*—Newm., Sta.
Eucosmia undulata.—Newm., Sta.
Cidaria truncata (russata).—Newm.
„ *testata.*—Newm.
„ *populata.*—Newm., Sta. In confinement.

Salix Caprea: (SALLOW).—89.
Vanessa polychloros.—Newm.
Apatura iris.—Buckl., Newm., Sta.
Smerinthus ocellatus.— Ent. xiii. 172, Buckl.
„ *populi.*—Sta.
Trochilium crabroniformis (bembeciformis).—Buckl., Sta., Ent. Mag. x. 161. In the wood.
Sarothripus undulanus (revayana).—Sta.
Ocneria dispar.—Ent. xviii. 243.
Dasychira pudibunda.—Ent. xxii. 151.
Orgyia antiqua —Buckl. In confinement.
Trichiura cratægi.—Sta.
Lasiocampa ilicifolia.—Sta.
Dicranura furcula.—Buckl., Newm., Sta.
„ *vinula.*—Buckl , Sta.
Pterostoma (Ptilodontis) palpina.—Buckl., Newm., Sta.
Notodonta dictæa.—Newm.
„ *dictæoides.*—Ent. xxi. 318.
„ *ziczac.*—Buckl., Newm., Sta.
Pygæra curtula.—Sta.
„ *anachoreta.*—Newm.
„ *pigra (reclusa).*—Newm., Sta.
Acronycta leporina.—Ent. xix. 133 ; xxii. 115.
„ *alni.*—Sta.

Salix Caprea—*continued.*
    Caradrina morpheus.—Ent. Mag. x. 254.
    Agrotis ashworthii.—Newm. In confinement.
    Tryphæna comes (orbona).—Newm. After hybernation.
    Noctua stigmatica (rhomboidea).—Newm. After hybernation.
        „ *brunnea.*—Newm.
        „ *triangulum.* In confinement.
        „ *festiva.*—Newm.
        „ *ditrapezium.* In confinement.
        „ *castanea v. neglecta.*—Ent. Mag. xxii. 242.
        „ *augur.*—Newm. After hybernation.
    Tæniocampa gothica.—Newm.
        „ *incerta (instabilis).*—Newm.
        „ *opima.*—Newm., Sta.
        „ *gracilis.*—Newm.
    Orthosia lota.—Newm.
    Cerastis vaccinii.—Newm.
    Xanthia fulvago (cerago).—Newm., Sta., Ent. Mag. vi. 262.
        In the catkins; afterwards on various low plants.
        „ *flavago (silago).*—Newm., Ent. Mag. vi. 262.
        „ *circellaris (ferruginea).*—Newm. On the buds.
    Tethea retusa.—Newm., Sta.
    Polia chi.—Newm., Ent. Mag. ix. 290.
    Cleoceris viminalis.—Newm.
    Aplecta nebulosa.—Newm. After hybernation.
        „ *occulta.*—Ent. Mag. xii. 66. In confinement.
    Hadena adusta.—Newm.
        „ *glauca.*—Newm.
        „ *rectilinea.*—Newm.
    Asteroscopus sphinx (cassinea).—Sta.
    Brephos notha.—Sta.
    Gonoptera libatrix.—Newm.
    Madopa salicalis.—Sta.
    Zonosoma (Ephyra) orbicularia.—Sta., Ent. x. 98.
    Nyssia zonaria.—Ent. Mag. xxiii. 61.
    Acidalia rubiginata (rubricata).
    Cabera exanthemaria.—Newm., Sta.
    Macaria alternata.—Newm., Sta.
        „ *notata.*—Newm., Sta.
    Numeria pulveraria.—Newm., Sta.
    Fidonia carbonaria.—Newm.
    Lomaspilis marginata.—Newm., Sta.
    Oporabia filigrammaria.—Newm.
    Eupithecia tenuiata.—Newm., Sta. In the catkins.
        „ *exiguata.*—Newm.
    Lobophora sexalisata (sexalata).—Newm.
        „ *halterata (hexapterata).*—Newm., Sta.
        „ *carpinata (lobulata).*—Newm., Sta.

Salix Caprea—*continued.*
   Hypsipetes ruberata.—Sta.
      „   *sordidata (elutata).* —Newm., Sta.
   Eucosmia undulata.—Newm., Sta.
   Cidaria truncata (russata).—Newm.
     „ *testata.*—Newm.
     „ *populata.*—Newm., Sta. In confinement.

Salix repens : DWARF SALLOW.—92.
   Callimorpha dominula.—Ent. xii. 79.
   Hadena contigua.—Ent. Mag. xvii. 171.
   Noctua c-nigrum.
     „ *triangulum.*
     „ *rubi.*
     „ *baia (baja).*
     „ *xanthographa.*

Populus : POPLAR.
   Apatura iris.—Sta.
   Smerinthus ocellatus.—Sta.
     „ *populi.*—Sta., Buckl.
   Trochilium apiformis.—Newm., Sta., Ent. Mag. x. 161. In the stems and wood.
   Cossus ligniperda.—Sta. In the wood.
   Zeuzera pyrina (æsculi).—Buckl. In the wood.
   Heterogenea asella (asellus).—Sta.
   Leucoma salicis.—Sta.
   Pœcilocampa populi.—Newm., Sta.
   Dicranura bifida.—Buckl., Sta.
     „ *vinula.*—Newm., Sta.
   Glyphisia crenata.—Newm., Sta.
   Pterostoma (Ptilodontis) palpina.—Buckl., Newm., Sta.
   Lophopteryx camelina.—Buckl.
   Notodonta trilophus.—Sta., Buckl.
     „ *ziczac.*—Newm., Sta.
   Pygæra curtula.—Newm., Sta.
   Acronycta megacephala.—Newm., Sta.
   Tæniocampa populeti.—Newm.
   Pachnobia rubricosa.—Ent. Mag. xxiii. 7.
   Orthosia upsilon.—Sta., Ent. Mag. ix. 248.
   Tethea subtusa.—Newm., Sta., Ent. Mag. ix. 248.
     „ *retusa.*—Sta.
   Catocala fraxini.—Sta.
     „ *nupta.*—Sta.
   Epione apiciaria.—Newm., Sta.
   Amphidasys betularia —Ent. xix. 253.
   Tephrosia crepuscularia.—Sta.

FOOD-PLANTS AND LARVÆ. 135

Populus nigra (pyramidalis) : LOMBARDY POPLAR.
   Smerinthus populi.—Newm.
   Trochilium crabroniformis (bembeciformis). In the bark and wood.
   Leucoma salicis.—Newm.

Populus alba : WHITE POPLAR.—60.
   Smerinthus ocellatus.—Ent. xviii. 181.

Populus nigra : BLACK POPLAR.
   Smerinthus ocellatus.—Buckl.
   Trochilium apiformis.—Buckl. In the bark and wood.
      „   *crabroniformis (bembeciformis).*—Ent. xx. 100. In the bark and wood.
   Sciopteron tabaniformis (vespiformis).—Sta., Ent. Mag. x. 161. In the stems and roots.
   Trichiura cratægi.—Ent. xv. 234.
   Bombyx quercus.—Ent. Mag. xix. 237.
   Dicranura vinula.—Buckl.
   Glyphisia crenata.—Buckl.
   Notodonta dictæa.—Newm.
   Pygæra anachoreta.—Newm.
   Cymatophora or.—Newm., Sta.

Populus tremula : ASPEN.—100.
   Vanessa polychloros.—Newm.
   Smerinthus ocellatus.—Ent. xiii. 108.
      „   *populi.*—Sta.
   Trochilium apiformis.—Newm., Ent. Mag. x. 161. In the wood.
   Sciopteron tabaniformis (asiliformis).—Newm., Sta. In the bark and wood.
   Drepana (Platypteryx) falcataria (falcula).—Sta.
   Dicranura bifida.—Buckl., Newm.
   Pterostoma (Ptilodontis) palpina.—Buckl.
   Notodonta dictæa.—Ent. xiii. 108., Buckl.
      „   *trilophus.*—Sta.
   Pygæra curtula.—Newm.
      „   *pigra (reclusa).*—Ent. xvi. 37.
   Cymatophora octogesima (ocularis).—Newm., Sta.
      „   *or.*—Newm.
   Tæniocampa populeti.—Newm., Ent. xii. 165.
      „   *munda.*—Sta.
   Xanthia circellaris (ferruginea).—Newm.
   Tethea subtusa.—Ent. xiii. 108.
   Brephos notha.—Newm., Sta., Ent. Mag. ix. 41.
   Catocala fraxini.—Sta.
   Lobophora halterata (hexapterata).—Newm., Sta.
   Cidaria silaceata.—Sta.
      „   *testata.*—Sta.

Empetrum nigrum : BLACK CROWBERRY.—71.
  Acidalia contiguaria. Ent. Mag. iii. 69. In confinement.

Juniperus communis: JUNIPER—73.
  Eupithecia helveticaria.—Newm., Sta.
  „   „   *v. arceuthata.*—Newm.
  „   *pusillata.*—Sta.
  „   *indigata.*—Newm.
  „   *sobrinata.*—Newm., Sta.
  Thera juniperata.—Newm., Sta.
  „   *simulata.*—Newm., Sta.

Taxus baccata : YEW.—48.
  Boarmia abietaria.—Ent. xxii. 287. In confinement.

Pinus : FIR.
  Sphinx pinastri.—Sta.
  Psilura monacha.—Sta.
  Boarmia abietaria.—Newm., Sta., Ent. xxii. 287.
  Tephrosia crepuscularia.—Ent. xix. 271.
  „   *biundularia.*—Ent. xix. 271.
  Macaria liturata.—Newm., Sta.
  Eupithecia indigata.—Sta.
  Thera firmata.—Newm., Sta.

Pinus sylvestris : SCOTCH FIR.—12.
  Panolis piniperda.—Newm., Sta.
  Ellopia prosapiaria (fasciaria).—Newm., Sta.
  Bupalus piniaria.—Newm., Sta.
  Thera variata.—Newm., Sta.

Pinus abies : SPRUCE FIR.
  Eupithecia lariciata.—Newm.
  „   *togata.*—Ent. xviii. 145., Ent. Mag. xii. 157.
  In the cones.

Larix Europæa.—LARCH.
  Tephrosia biundularia.—Sta.
  Eupithecia lariciata.—Newm.

Listera ovata : TWAYBLADE —102.
  Spilosoma fuliginosa.—Ent. Mag. xxi. 158.

Iris Pseudacorus : YELLOW IRIS.—112.
  Spilosoma urticæ.—Ent. Mag. xxi. 207.
  Nonagria sparganii.—Ent. xiii. 49. In the stems of this and other *Iris.*
  Apamea leucostigma (fibrosa).—Newm., Sta. In the stems.
  „   *ophiogramma.* In the stems.
  Calocampa vetusta.
  Plusia festucæ.—Ent. xix. 209.

Scilla nutans: SQUILL ; BLUEBELL.—109.
    Noctua umbrosa.—Ent. xviii. 210. On the seeds.

Juncus: RUSH.
    Erebia epiphron (cassiope).—Newm.

Juncus lamprocarpus.—109.
    Cœnobia rufa (despecta). In the stems.

Luzula vernalis (pilosa): BROAD-LEAVED WOOD-RUSH. 106.
    Hesperia sylvanus.—Buckl.
    Leucania turca.—Newm., Sta.
    Xylophasia scolopacina.—Newm.
    Herminia cribralis.—Ent. Mag. x. 103.

Luzula campestris: FIELD WOOD-RUSH ; CUCKOO-GRASS. 105.
    Leucania turca.—Newm., Sta.
    Xylophasia scolopacina.—Newm. (?)

Luzula maxima: WOOD-RUSH.—105.
    Triphæna fimbria.—Ent. xiv. 155.

Typha latifolia: REED MACE.—79.
    Nonagria cannæ.—Newm., Sta.
    ,,   *sparganii.*—Ent. xiii. 51. In the stems.
    ,,   *arundinis (typhæ).*—Newm., Sta. In the stems.
    ,,   *brevilinea.*

Typha angustifolia.—57.
    Nonagria sparganii.—In the stems.

Sparganium ramosum: BUR-REED.—18.
    Nonagria sparganii.—Ent. xiii. 50. In the stems.

Sparganium simplex: UNBRANCHED BUR-REED.—92.
    Plusia festucæ.—Sta.

Cyperus longus: CYPRESS-ROOT.—7.
    Hydrœcia micacea.—Newm., Sta.

Scirpus.
    Nonagria brevilinea.—Ent. x. 17. In the stems.

Scirpus cæspitosus.—101.
    Xylophasia scolopacina.—Sta. Also on other *Scirpi.*

Rynchospora alba: WHITE BEAK-RUSH.—75.
    Cænonympha typhon (davus).—Buckl.
    ,,   ,,   ,,   *v. rothliebi.*—Newm.

Cladium germanicum: TWIG-RUSH.—37.
    Lælia cœnosa.—Sta., Ent. xii. 79.
    Apamea leucostigma (fibrosa).—Ent. Mag. xx. 176.

Carex : SEDGE.
   Lælia cænosa.—Ent. xiii. 171.
   Arsilonche albevenosa (venosa).—Ent. xiii. 172.
   Leucania impura.—Sta.
   Tapinostola fulva.—Newm., Sta. In the stems.
   Hydrœcia micacea.—Newm. At the roots.
   Hydrelia uncula (unca).—Sta.
   Plusia festucæ.—Sta.
   Apamea unanimis.

Carex glauca.—106.
   Phothedes captiuncula.—Ent. Mag. xviii. 76.

Carex sylvatica.—83.
   Leucania littoralis.—Newm. In confinement.
   Hydrelia uncula (unca).—Ent. Mag. vi. 232. In confinement.
   Herminia cribralis.—Ent. x. 103.

Carex paludosa.—70.
   Tapinostola fulva.—Ent. Mag. xvii. 114. In the stems.

Carex riparia.—72.
   Leucania littoralis.—Newm.

Eriophorum angustifolium : COMMON COTTON GRASS.—109.
   Cænonympha typhon (davus).—Buckl.
   Tapinostola fulva.—Ent. xvi. 261. In the stems of this and other *Eriophora*.
   Celæna Haworthii.—Newm., Sta. In the stems of this and other *Eriophora*.

Eriophorum vaginatum : HARE'S-TAIL COTTON GRASS.—89.
   Celæna hawoithii.—Ent. Mag. ix. 195.

Phalaris canariensis : CANARY GRASS.
   Apamea unanimis.
      „   *ophiogramma.*

Phalaris arundinacea : REED CANARY GRASS.—109.
   Leucania straminea.—Ent. Mag. viii. 249.
   Phothedes captiuncula.—Ent. Mag. xviii. 76.
   Apamea ophiogramma.
      „   *gemina.*—Ent. Mag. x. 275. In confinement.
      „   *unanimis.*

Milium effusum : MILLET.—80.
   Epinephele hyperanthes.—Newm., Sta.

Phleum pratense : TIMOTHY GRASS.—104.
   Melanargia galatea.—Sta.

Agrostis canina.—93.
   Erebia æthiops (medea).—Newm.

Calamagrostis epigeios.—57.
  Hesperia actæon.—Newm., Sta.
Ammophila arundinacea: MARRAM.—60.
  Satyrus semele.—Ent. Mag. ix. 21.
  Leucania littoralis.—Newm., Ent. Mag. ix. 21.
  Tæniocampa opima.—Ent. Mag. ix. 21.
Aira caryophyllea: HAIR-GRASS.—105.
  Cerigo matura (cytherea).—(?).
Aira præcox: EARLY HAIR-GRASS.—108.
  Erebia epiphron (cassiope).—Buckl.
    „   *æthiops (medea)* —Buckl., Ent. Mag. vii. 65.
  Satyrus semele.—Buckl., Newm.
  Eremobia ochroleuca.—(?).
Corynephorus (Aira) canescens.—4.
  Mamestra furva.—Newm., Sta.
Deschampsia (Aira) cæspitosa: TUSSOCK-GRASS.—111.
  Erebia æthiops (medea).—Buckl., Ent. Mag. vii. 65.
  Satyrus semele.—Buckl., Newm.
  Epinephele hyperanthes.—Newm.
  Leucania pallens.—Ent. Mag. iii. 68.
  Miana arcuosa.—Newm., Sta., Ent. Mag. ix. 248 ; vii. 88 ;
         vii. 260. In the stems and roots.
    „ *fasciuncula.*
Deschampsia (Aira) flexuosa.—105.
  Erebia epiphron (cassiope).—Buckl.
Holcus lanatus: MEADOW SOFT-GRASS.—111.
  Hesperia thaumas (linea).—Buckl.
    „   *sylvanus.*—Newm., Sta.
  Agrotis vestigialis (valligera).—(?). On the roots.
    „ *suffusa.*—(?). On the roots.
  Anchocelis lunosa.—(?).
Arrhenatherum avenaceum.—110.
  Tapinostola bondii.—Ent. xi. 252.
  Miana bicoloria (furuncula).—Ent. xi. 252.
Phragmites communis: REED.—104.
  Macrogasta castaneæ (arundinis).—Buckl., Newm., Sta. In
      the stems.
  Lælia cænosa.—Newm., Sta.
  Arsilonche albevenosa (venosa).—Newm.
  Leucania obsoleta.—Newm., Sta.
    „   *impudens (pudorina).*—Newm.
    „   *straminea.*—Ent. Mag. viii. 249.
  Meliana flammea.—Ent. Mag. xx. 64. In the stems.
  Senta maritima (ulvæ).—Newm., Sta. In the stems.

Phragmites communis—*continued.*
   Calamia phragmitidis.—Newm., Sta. In the stems.
   Tapinostola helmanni.
   Nonagria geminipuncta.—Newm., Sta., Ent. Mag. x. 231.
       In the stems.
     "   *neurica.*—Sta., Ent. Mag. x. 275. In the stems.
     "   *brevilinea.*—Ent. Mag. xxii. 272. In the stems.
     "   *lutosa.*—Newm., Sta. In the stems.

Cynosurus cristatus: DOG'S-TAIL.—112.
   Cænonympha pamphilus.—Sta.

Molinia cœrulea: PURPLE MELIC-GRASS.—106.
   Erastria fasciana (fuscula).—Ent. Mag. xi. 66.

Dactylis glomerata: COCK'S-FOOT GRASS.—112.
   Melanargia galatea.—Buckl.
   Pararge megæra.—Newm.
     "   *egeria.*—Buckl.
   Epinephele tithonus.—Buckl.
   Hesperia sylvanus.—Buckl.
   Leucania putrescens.—(?).
     "   *comma.*—Newm.
     "   *pallens.*—Ent. Mag. iii. 68.
   Miana strigilis.—Ent. xvi. 91. In the stems.
   Tryphæna orbona (subsequa).—Ent. Mag. xvii. 211; Ent. x. 48.
   Eremobia (Ilarus) ochroleuca.—Ent. Mag. viii. 21. In confinement; on the seeds.

Briza media: QUAKING GRASS; MAIDENHAIR.—105.
   Xylophasia scolopacina.—Sta.

Briza minor: SMALL QUAKING GRASS.—7.
   Xylophasia scolopacina.—Sta.

Poa annua: ANNUAL MEADOW-GRASS.—110.
   Erebia epiphron (cassiope).—Buckl., Newm.
     "   *æthiops (medea).*—Newm. Also on several species of *Poa.*
   Epinephele tithonus—Buckl. Sta.
     "   *hyperanthes.*—Newm., Sta.
   Cænonympha pamphilus.—Sta.
   Odonestis potatoria.
   Leucania pallens.
   Hydrœcia nictitans. On the roots.
   Xylophasia lithoxylea. On the roots.
     "   *sublustris.* On the roots.
     "   *monoglypha (polyodon).* On the roots.
   Aporophyla australis.—Ent. Mag. vi. 13.
   Neuronia popularis.

Poa annua—*continued.*
   Heliophobus hispidus.—Ent. xiv. 135.
   Cerigo matura (cytherea).—Ent. xiii. 141.
   Luperina testacea.—On the stems.
      „   *cespitis.*
   Charæas graminis. On the roots.
   Pachetra leucophæa.—Ent. xv. 162., Ent. Mag. xix. 43.
      "In tufts of grass growing in woods" and on commons.
   Noctua umbrosa.
      „   *xanthographa.*
   Apamea didyma (oculea).
      „   *gemina.*—Ent. Mag. x. 275. In confinement.
   Epunda lutulenta.—Ent. Mag. vi. 235. In confinement.
   Aplecta tincta.
   Crymodes exulis.—Newm. On this and other species of
      Poa.
   Stilbia anomala.—Ent. xviii. 3.
   Bankia argentula.—Ent. Mag. xx. 77 ; xxiii. 5.
   Camptogramma bilineata.

Poa nemoralis : WOOD MEADOW-GRASS.—82.
   Pachetra leucophæa.—Ent. xiii. 234.

Poa pratensis.—108.
   Erebia æthiops (medea).
   Epinephele janira.—Sta.
   Bankia argentula.—Ent. Mag. xxiii. 5.
   Eubolia limitata (mensuraria).

Poa trivialis.—108.
   Leucania straminea.—(?).

Glyceria aquatica.—76.
   Arsilonche albevenosa (venosa).—Sta.
   Tapinostola fulva.—Newm., Sta.
   Bankia argentula.—Ent. Mag. xxiii. 5.

Glyceria maritima.—67.
   Hydræcia nictitans.—Ent. Mag. xviii. 195.
   Rivula sericealis.—Ent. Mag. xix. 49.

Festuca ovina : SHEEP'S FESCUE-GRASS.—110.
   Erebia epiphron (cassiope).—Newm.
  * *Emydia striata (grammica).*—Sta.
   Pachetra leucophæa.—(?).
   Agrotis ashworthii.—Newm., Ent. xxiii. 6.
   Acidalia ochrata.—(?). Sta.

Brachypodium pinnatum.—35.
   Rivula sericealis.—Ent. Mag. xix. 50.

Brachypodium sylvaticum.—105.
   Hesperia actæon.—Buckl., Ent. Mag. x. 86.
   Hesperia thaumas (linea).—Buckl., Ent. Mag. xix. 244.
   Carterocephalus palæmon (paniscus).—Buckl.
   Rivula sericealis.—Ent. Mag. xix. 50.

Lolium : DARNEL.
   Eremobia (Ilarus) ochroleuca.—Ent. ix. 18.

Agropyron (Triticum) : WHEAT.
   Apamea basilinea.—Newm., Sta. On the grain.

Agropyron (Triticum) repens : COUCH GRASS.—108.
   Pararge egeria.—Sta.
   Satyrus semele.—Buckl., Newm.
   Epinephele tithonus.—Newm.
      „   *hyperanthes.*—Buckl., Newm.
   Hesperia actæon.—Buckl., Ent. Mag. x. 86.
      „   *sylvanus.*—Buckl.
   Leucania conigera.—Newm., Ent. Mag. iii. 137.
      „   *pallens,*—Ent. Mag. iii. 68.
      „   *albipuncta.*—(?).
   Apamea gemina.—Ent. Mag. x. 275. In confinement.
      „   *unanimis.*—Ent. Mag. viii. 207. In confinement.
   Tryphæna orbona (subsequa).—Ent. Mag. xvii. 211., Ent. x. 48.

Agropyron (Triticum) pungens.
   Hesperia actæon.—Ent. Mag. x. 86. In confinement.

Agropyron (Triticum) junceum.—36.
   Hesperia actæon.—Buckl. (?).
   Agrotis cursoria.—Ent. Mag. ix. 14.

Nardus stricta : MAT-GRASS.—107.
   Erebia epiphron (cassiope).—Newm.
   Cænonympha pamphilus.—Newm.
   Leucania vitellina.—(?).

Elymus arenarius : LYME-GRASS.—30.
   Tapinostola elymi.—Ent. Mag. viii. 69. In the stems and roots.

Pteris aquilina : BRAKE ; BRACKEN.—112.
   Hepialus velleda.—Buckl., Newm., Sta., Ent. Mag. vii. 84. On the roots.
      „   *hectus.*—Buckl., Ent. Mag. v. 177. On the roots.
   Euplexia lucipara.—Newm.
   Hadena pisi.—Newm.
   Panagra petraria.—Newm.

Lastræa (nephrodium) felix-mas : MALE FERN.—112.
   Euplexia lucipara.—Newm.

Lichen.
   * *Nacilia ancilla.*—Newm. On trees.
   Nudaria mundana.—Newm., Sta. On trees. Buckl., Ent. Mag. viii. 171. On walls.
       ,, *senex.*—Buckl., Ent. Mag. viii. 170.
   Setina irrorella.—Sta., Buckl., Ent. Mag. viii. 171. On stones on the sea-coast above high water-mark.
   Calligenia miniata.—Newm., Sta. On oaks. Sta. On beech and birch.
   Lithosia mesomella.—Buckl., Sta., Ent. Mag. viii. 172. On oaks. In confinement on green or withering sallow-leaves.
       ,, *muscerda.*—Buckl. Probably on sallows in wet places.
       ,, *sororcula (aureola).*—Buckl., Ent. Mag. v. 113. On oaks. Newm. On larch. Ent. Mag. xi. 158. On beeches. Sta. On fir and pine.
       ,, *lutarella (pygmæola).*—Buckl.
       ,, *griseola.*—Buckl. On mosses and withered sallow-leaves in confinement.
       ,,   ,, *v. stramineola.*
       ,, *deplana.*—Buckl., Ent. Mag. v. 112. On yews. Newm. On oaks. Sta. On oaks and beech.
       ,, *lurideola (complanula).*—Newm. On oak, blackthorn and larch. Sta. On poplars and walls.
       ,, *complana.*—Buckl. On firs. Newm., Sta. On blackthorn and firs.
       ,, *sericea (molybdeola).*—Buckl., Newm. Probably on heath or stones under. In confinement on withered oak or sallow-leaves.
   Gnophria quadra.—Newm., Sta. On trees.
       ,, *rubricollis.*—Buckl. On beech. Newm., Sta. Various trees.
   Bryophila muralis (glandifera).—Newm., Sta. On walls and buildings.
       ,, *par.*—On walls and building.
       ,, *algæ.*—Newm. On trees.
       ,, *perla.*—Newm., Sta. On walls.
   Aventia flexula.—Sta., Ent. Mag. x. 42. On thorns, cherry and yew.
   Cleora glabraria.—Newm., Sta. On firs. Ent. Mag. xii. 84. On oaks.
       ,, *lichenaria.*—Newm., Sta. On trees.
       ,, *angularia (viduaria).*
   * *Tephrona sepiaria (cineraria).*—Newm., Sta. On walls.

Equisetum arvense : FIELD MARE'S-TAIL.—108.
    Hydræcia micacea.—Ent. Mag. vi. 164. In the stems.

Equisetum limosum.—103.
    Hydræcia micacea.—Ent. Mag. vi. 164. In the stems.

Lichen caninus.
    Calligenia miniata.—Buckl., Ent. Mag. v. 111. On oaks. In confinement on withered oak and sallow leaves.
    Lithosia griseola v. stramineola.—Buckl., Ent. Mag. v. 110.
    Gnophria quadra.—Buckl. On oaks.
    Nudaria senex.—Buckl., Ent. Mag. viii. 170. In confinement on this and decayed sallow or bramble-leaves.
    Nola strigula.—Buckl. On oaks.

Fungi.
    Boletobia fuliginaria.—Sta. On rotten wood.

Dried Plants.
    * *Acidalia herbariata.*—Sta.

INDEX.

	PAGE		PAGE
A		Annual Meadow-Grass	140
		Anthemis	105
Abraxas (Zerene)	66	*Anthriscus*	100
Acacia	92	*Anthyllis*	91
Acer	88	*Anticlea*	75
Acherontia	8	*Apamea*	31
Achillea	105	*Apatura*	5
Acidalia	60	*Aplasta*	65
Acontia	52	*Aplecta*	46
Acosmetia	33	*Aporia*	1
Acronycta	24	*Aporophyla*	30
Æsculus	88	Apple	97
Agriopis	45	*Arabis*	82
Agrophila	52	*Arctia*	16
Agropyron	142	*Arctium*	107
Agrostis	138	*Arctostaphylos*	109
Agrotis	33	*Arenaria*	86
Aira	139	„ (*Cherleria*)	86
Alchemilla	96	*Argynnis*	3
Alder	125	*Armeria*	111
Alder Buckthorn	88	*Arrhenatherum*	139
Aleucis	64	*Arsilonche*	26
Alnus	125	*Artemisia*	105
Alpine Lady's Mantle	96	Ash	111
Alternate-leaved Golden Saxifrage	99	Aspen	135
		Asperula	103
Ammophila	139	*Asphalia*	24
Amphidasys	57	*Aspilates*	66
Amphipyra	39	*Aster*	104
Anagallis	111	*Asteroscopus*	49
Anaitis	79	*Asthena*	60
Anarta	51	*Atriplex*	118
Anchocelis	41	*Atropa*	113
Anemone	81	Autumnal Gentian	112
Angelica	101	Avens, Common	95
Angerona	55	*Aventia*	53
Anisopteryx	67	*Axylia*	29

B.

	PAGE
Ballota	116
Bankia	52
Bapta (corycia)	64
Barbarea	82
Barbary	81
Bartsia	114
Bastard Alkanet	113
Beech	129
Beet	117
Berberis	81
Beta	117
Betula	123
Bilberry	109
Birch	123
Bird's-foot	92
Bird's-foot Trefoil	91
Biston	57
Biting Stonecrop	99
Black Crow-berry	136
„ Currant	99
„ Horehound	116
„ Medick	89
„ Mullein	113
„ Poplar	135
Blackthorn	92
Bladder Senna	92
Blue Bell	137
Boarmia	58
Bog Whortleberry	109
Boletobia	53
Bombyx	20
Bomolocha	54
Borage	113
Borago	113
Bracken	142
Brachypodium	141
Brake	142
Bramble	94
Brassica	82
Brephos	54
Briza	140
Broad-leaved Dock	120
„ „ Wood-rush	137
Broom	89

	PAGE
Bryophila	24
Buck-bean	112
Buck's-horn Plantain	117
Buckthorn	88
Bulbous Buttercup	81
Bupalus	65
Burdock	107
Burnet Rose	96
Burnet Saxifrage, Common	100
Bur-reed	137
Butter Bur	106

C.

	PAGE
Cabera	64
Cakile	83
Calamagrostis	139
Calamia	27
Calligenia	14
Callimorpha	15
Calluna	109
Calocampa	48
Calymnia	42
Calystegia	113
Campanula	108
Camptogramma	76
Canary Grass	138
Canterbury Bells	108
Caradrina	32
Cardamine	82
Carduus	107
Carex	138
Carpinus	126
Carsia	79
Carterocephalus	8
Castanea	129
Catephia	53
Catocala	53
Celæna	32
Centaurea	107
Cerastis (Glæa)	41
Cerastium	85
Cerigo	30
Chærophyllum	100
Chamomile	105

INDEX. 147

	PAGE		PAGE
Chararas	30	*Cotyledon*	99
Chariclea	51	Couch Grass	142
Cheimatobia	67	Cow-berry	109
Chenopodium	117	Cow Parsnep	101
Cherlock	82	Cowslip	111
Chesias	79	Cow-wheat	114
Chestnut	129	Crack Willow	130
Chicory	107	Cranberry	109
Chickweed	86	*Cratægus*	97
Chærocampa	9	Creeping Crowfoot	81
Chrysanthemum	105	*Crepis*	107
Chrysosplenium	99	*Crocallis*	56
Cidaria	77	Cross-leaved Heath	110
Cilix	21	*Crymodes*	46
Circæa	100	Cuckoo-flower	82
Cirrhædia	42	,, grass	137
Cladium	137	*Cucullia*	49
Clematis	81	Cud-weed, Common	105
Cichorium	107	Curled Dock	120
Cleoceris	45	*Cymatophora*	24
Cleora	57	*Cynoglossum*	112
Cloantha	30	*Cynosurus*	140
Clustered Bell-flower	108	*Cyperus*	137
Cnicus	107	Cypress-root	137
Cock's-foot Grass	140	Cypress Spurge	121
Cnobia	28	*Cytisus (Sarothamnus)*	89
Cænonympha (Chortobius)	6		
Cælias	2		
Collix	74	**D.**	
Colt's-foot	106		
Colutea	92	*Dactylis*	140
Comfrey	113	Dame's Violet	82
Conopodium	100	*Danais*	4
Convolvulus	113	Dandelion	108
Coremia	76	Darnel	142
Corn Chamomile	105	*Dasycampa*	41
,, Cockle	85	*Dasychira*	18
,, Feverfew	105	*Dasodia*	59
,, Gromwell	113	*Dasypolia*	45
,, Sow-thistle	108	*Daucus*	101
Cornus	101	Deadly Nightshade	113
Corylus	126	*Deilephila*	9
Corynephorus (Aira)	139	*Deiopeia*	15
Cosmia (Euperia)	42	*Demas*	24
Cossus	17	Deptford Pink	84
Cotton Grass, Common	138	*Deschampsia (Aira)*	139

L 2

148 LARVA COLLECTING AND BREEDING.

	PAGE		PAGE
Devil's-bit Scabious	104	Epione	54
Dewberry	95	Epunda	45
Dianthœcia	43	Equisetum	144
Dianthus	84	Erastria	52
Dicycla	42	Erebia	5
Dicranura (Cerura)	21	Eremobia (Ilarus)	43
Digitalis	114	Erica	110
Diloba	26	Eriogaster	19
Dipsacus	104	Eriophorum	138
Dipterygia	30	Erodium	87
Dock	119	Erysimum	82
Dog Rose	96	Eubolia	78
„ Violet	83	Euchelia	15
Dog-wood	101	Enchloë	2
Dog's Chamomile	105	Euclidia	52
„ Mercury	121	Eucosmia	77
Dog's-tail Grass	140	Eugonia (Ennomos)	56
Do-not-touch-me	87	Euonymus	87
Drepana (Platypteryx)	21	Eupa'orium	104
Dried Plants	144	Euphorbia	121
Dropwort	94	Euphrasia	114
Dutch Clover	90	Eupisteria	60
Dyer's Green-weed	89	Eupithecia	69
„ Rocket	83	Euplexia	45
Dwarf Sallow	134	Eurymene	55
		Evening Campion	85
		Everlasting Pea	92
E.		Eye-bright	114
Earias	13		
Early Hair-grass	139	F.	
„ Winter Cress	82		
Echium	113	Fagus	129
Elder	101	Festuca	141
Ellopia	55	Feverfew, Common	105
Elm	121	Fiddle Dock	120
Elymus	142	Fidonia	65
Ematurga	65	Field Bindweed	113
Emmelesia	68	„ Forget-me-not	113
Empetrum	136	„ Gentian	112
Emydia	15	„ Mare's-tail	144
Enchanter's Nightshade	100	„ Mouse-ear Chickweed	85
Endromis	20		
English Catchfly	84	„ Scabious	104
Epilobium	100	„ Thistle	107
Epinephele	5	„ Wood-rush	137

	PAGE		PAGE
Field Wormwood	106	Ground Ivy	115
Filago	105	Groundsel	106
Fir	136	Guelder Rose	101
Flixweed	82		
Flowering Currant	99		
" Willow	100	**H.**	
Forget-me-not	113		
Foxglove	114	*Habrostola*	50
Fragaria	95	*Hadena*	46
Fraxinus	111	Hair-grass	139
Fungi	144	Hairy Mint	115
Furze	89	*Halia*	64
		Harebell	109
		Hare's Parsley	100
G.		Hare's-tail Cotton-grass	138
		Hazel	126
Galeopsis	115	Heartsease	83
Galium	102	Heath Bedstraw	103
Genista	89	Heather	109
Gentiana	112	*Hecatera*	44
Geometra	59	*Hedera*	101
Germander Speedwell	114	Hedge Bedstraw	102
Geum	95	" Mustard, Common	82
Glyceria	141	" Wound-wort	115
Glyphasia	22	*Heliaca* (*Heliodes*)	51
Gnophos	59	*Helianthemum*	83
Gnophria	15	*Heliophobus*	30
Golden-rod	104	*Heliothis*	51
Gonoptera	50	*Helleborus*	81
Gonopteryx	2	*Hemerophila*	57
Good King Henry	117	*Hemithea*	60
Gooseberry	99	Hemlock Stork's-bill	87
Goose Grass	103	Hemp Agrimony	104
Gorse	89	" Nettle	115
Gortyna	29	Henbane	113
Grammesia	32	*Hepialus*	17
Great Bindweed	113	*Heracleum*	101
" Burnet Saxifrage	100	*Herminia*	53
" Hairy Willow-herb	100	*Hesperia*	8
" Wild Valerian	103	*Hesperis*	82
Greater Bird's-foot Trefoil	91	*Heterogenea*	18
" Periwinkle	112	*Hieracium*	107
" Plantain	116	*Himera*	56
" Stitchwort	85	*Hippocrepis*	92
		Hoary Ragwort	106

					PAGE					PAGE
Holcus	.	.	.	.	139	Knapweed	.	.	.	107
Holly	.	.	.	.	87	Knot-grass	.	.	.	113
Honeysuckle	.	.	.	101	Knotted Fig-wort	.	.	114
Hop	.	.	.	.	122					
„ Trefoil	.	.	.	91					
Hornbeam	.	.	.	126	**L.**				
Horse Chestnut	.	.	88						
Horse-shoe Vetch	.	.	92	Lactuca	.	.	.	.	108
Hound's-tongue	.	.	112	Lady's Bedstraw	.	.	102
Humulus	.	.	.	122	„ Fingers	.	.	91
Hybernia	.	.	.	66	Lælia	.	.	.	.	18
Hydrelia	.	.	.	52	Laphygma	.	.	.	30
Hydrilla	.	.	.	33	Lamium	.	.	.	115
Hydrœcia	.	.	.	29	Larch	.	.	.	.	136
Hylophila (Halias)	.	.	13	Larentia	.	.	.	68
Hyoscyamus	.	.	.	113	Laria	.	.	.	.	18
Hypena	.	.	.	.	54	Larix	.	.	.	.	136
Hypenodes	.	.	.	54	Lasiocampa	.	.	.	20
Hypericum	.	.	.	86	Lastræa (Nephrodium)	.	143
Hypsipetes	.	.	.	74	Lathyrus (Orobus)	.	.	92
Hyria	.	.	.	.	60	Leontodon	.	.	.	107
						Lepigonum (Spergularia)	.	86
						Leucania	.	.	.	26
I.						Leucoma	.	.	.	18
						Leucophasia	.	.	.	2
Ilex	.	.	.	.	87	Lichen	.	.	.	.	143
Impatiens	.	.	.	87	Ligdia	.	.	.	.	66
Ino	.	.	.	.	12	Ligustrum	.	.	.	112
Iodis	.	.	.	.	59	Lilac	.	.	.	.	112
Iris	.	.	.	.	136	Lime	.	.	.	.	87
Ivy	.	.	.	.	101	Limenitis	.	.	.	4
Ivy-leaved Toad-flax	.	114	Linaria	.	.	.	.	114
						Ling	.	.	.	.	109
						Listera	.	.	.	.	136
J.						Lithosia	.	.	.	.	14
						Lithospermum	.	.	113
Jack-by-the-hedge	.	82	Lithostege	.	.	.	79
Jasione	.	.	.	.	108	Lobophora	.	.	.	74
Juncus	.	.	.	.	137	Lolium	.	.	.	.	142
Juniper	.	.	.	.	136	Lomaspilis	.	.	.	66
Juniperus	.	.	.	136	Lombardy Poplar	.	.	135
						Lonicera	.	.	.	101
						Loosestrife	.	.	.	111
K.						Lophopteryx	.	.	.	22
						Lotus	.	.	.	.	91
Kidney Vetch	.	.	91	Lucerne	.	.	.	.	89

INDEX

	PAGE		PAGE
Luperina	30	*Melilotus*	90
Luzula	137	*Melitæa*	3
Lycæna	6	*Mentha*	115
Lychnis	85	*Menyanthes*	112
Lycium	113	*Menziesia*	111
Lyme-grass	142	*Mercurialis*	121
Lysimachia	111	*Mesotype*	79
Lythria	66	*Metrocampa*	55
Lythrum	100	*Miana*	32
		Milfoil	105
		Milium	138
M.		Milk Parsley	101
		Milk-wort, Common	84
Macaria	64	Millet	138
Macrogaster	17	*Minoa*	65
Macroglossa	10	*Miselia*	45
Madder	102	*Molinia*	140
Madopa	54	*Moma* (*Diphthera*)	24
Maidenhair	140	Moss Campion	84
Male Fern	143	Mossy Cyphel	86
Mallow	87	„ Saxifrage	99
Malva	87	Moth Mullein	114
Mamestra	31	Mountain Ash	96
Mania	39	Mouse-ear Chickweed	85
Maple	88	„ „ Hawk-weed	107
Marjoram	115	Mugwort	105
Marram	139	Mullein	113
Marsh Flea-wort	106	Musk-thistle	107
„ Red Rattle	114	*Myosotis*	113
„ Sow-thistle	108	*Myrica*	123
„ Thistle	107		
„ Trefoil	112		
Mat-grass	142	**N.**	
Matricaria	105		
Meadow Crowfoot	81	*Nacilla*	13
„ Saxifrage	100	*Nardus*	142
„ Soft-grass	139	Narrow-leaved Bitter Cress	82
„ -sweet	94	„ „ Plantain	117
„ Vetchling	92	*Nasturtium*	82
Medicago	89	Needle Furze	89
Melampyrum	114	*Nemeobins*	7
Melanargia	5	*Nemeophila*	16
Melanippe	75	*Nemoria*	59
Melanthia	75	*Nepeta*	115
Meliana	28	*Neuria*	30
Melilot	90	*Neuronia*	30

152 LARVA COLLECTING AND BREEDING.

	PAGE		PAGE
Nisioniades	8	Pelurga	78
Noctua	35	Penagra	65
Nola	13	Pericallia	55
Nonagria	28	Perforated St. John's Wort	86
Notodonta	22	Persicaria, Common	119
Nottingham Catchfly	85	Petasites	106
Nudaria	14	Petty Spurge	121
Numeria	65	Peucedanum	101
Nyssia	57	Phalaris	138
		Phalera	23
		Phibalapteryx	76
O.		Phigalia	57
		Phleum	138
Oak	126	Phlogophora	45
Ocneria	18	Phorodesma	59
Odonestis	20	Phothedes	32
Odontopera	56	Phragmites	139
Onobrychis	92	Phytometra	52
Ononis	89	Pig-nut, Common	100
Onopordon	107	Pieris	1
Ophiodes	52	Pilewort	81
Oporabia	67	Pimpinella	100
Oporina	41	Pinus	136
Orach	118	Plantago	116
Orgyia	19	Plantain	116
Origanum	115	Plum	93
Ornithopus	92	Plusia	50
Orpine	99	Poa	140
Orthosia	40	Pœcilocampa	19
Osier	131	Polia	44
		Polygala	84
		Polygonum	118
P.		Polyommatus	6
		Poplar	134
Pachetra	30	Populus	134
Pachnobia	39	Portland Spurge	121
Pachycnemia	66	Porthesia (Liparis)	18
Panolis	39	Potentilla	95
Papaver	81	Poterium	96
Papilio	1	Prickly Glass-wort	118
Pararge	5	Primrose	111
Parietaria	123	Primula	111
Pear	96	Privet	112
Pechypogon	54	Prunella	115
Pedicularis	114	Prunus	92
Pellitory of-the-wall	123	Pseudoterpna	59

INDEX

	PAGE		PAGE
Psilura	18	*Rosa*	96
Psodos	59	Rough Chervil	100
Pteris	142	„ Hawk-bit	107
Pterostoma (Ptilodontis)	22	*Rubia*	102
Ptilophora	22	*Rubus*	94
Purple Clover	90	*Rumex*	119
„ Dead-nettle	115	*Rumia*	55
„ Heath, Common	110	Rush	137
„ Loosestrife	100	*Rusina*	133
„ Melic-grass	140	*Rynchospora*	137
„ Sandwort	86		
Pygæra (Clostera)	23		
Pyrus	96		

S.

		Saintfoin	92
Q.		Salad Burnet, Common	96
		Salix	129
Quaking Grass	140	Sallow	131
Quercus	126	*Salsola*	118
		Sambucus	101
		Sandwort	86
R.		*Sarothripus*	13
		Saturnia	21
Ragwort	106	*Satyrus*	5
Ragged Robin	85	*Saxifraga*	99
Ranunculus	81	*Scabiosa*	104
Rape	82	Scarlet Pimpernel	111
Raspberry	95	*Scilla*	137
Red Bartsia	114	*Sciopteron*	11
„ Bear-berry	109	*Scirpus*	137
„ Campion	85	*Scodiona*	65
„ Currant	99	*Scopelosema*	41
„ Hemp-nettle	115	*Scoria*	65
„ Poppy	81	Scotch Cinquefoil	95
Reed	139	„ Fir	136
„ Canary Grass	138	„ Thistle	107
„ Mace	137	Scottish Menziesia	111
Reseda	83	*Scotosia*	77
Rest-harrow	89	*Scrophularia*	114
Rhamnus	88	Sea Campion	84
Rhinanthus	114	„ Orach	118
Ribes	99	„ Plantain	117
Rivula	53	„ Rocket	83
Robinia	92	„ Sandwort	86
Rock Cress	82	„ Spurge	121
„ Rose	83	„ Star-wort	104

	PAGE		PAGE
Sea Wormwood	106	Strawberry-leaved Cinquefoil	95
Sedge	138	*Strenia*	64
Sedum	99	Sweet Gale	123
Selenia	55	„ Violet	83
Self-heal	115	„ Woodruff	103
Selidosema	65	Sycamore	88
Senecio	106	*Symphytum*	113
Senta	28	*Synia*	26
Sesia	11	*Syntomis*	13
Setina	14	*Syrichthus*	8
Sheep's Fescue-grass	141	*Syringa*	112
„ Scabious	108		
„ Sorrel	121		
Silaus	100	**T.**	
Silene	84		
Silver-weed	96		
Sisymbrium	82	*Tæniocampa*	39
Sleepwort	108	Tamarisk	86
Small Nettle	123	*Tamarix*	86
„ Quaking Grass	140	*Tanacetum*	105
„ Scabious	104	*Tanagra*	79
„ Teasel	104	Tansy	105
Smerinthus	10	*Tapinostola*	28
Smooth Hawk's-beard	107	*Taraxacum*	108
Solanum	113	*Taxus*	136
Solidago	104	Tea-tree	113
Sonchus	108	*Tephrona*	59
Sorrel	120	*Tephrosia*	58
Sow-thistle	108	*Tethea*	42
Spanish Catchfly	85	*Teucrium*	116
Sparganium	137	*Thalictrum*	81
Spear Thistle	107	*Thalpochares (Micra)*	52
Sphinx	8	*Thecla*	6
Spilosoma	16	*Thera*	74
Spindle	87	*Tholomiges (Schrankia)*	54
Spiræa	94	Three-stamened Willow	130
Spruce Fir	136	Thrift	111
Squill	137	*Thyatira*	24
Stachys	115	Thyme	115
Stauropus	22	Thyme-leaved Speedwell	114
Stellaria	85	*Thymus*	115
Sterrha	65	*Tilia*	87
Stilbia	32	*Timandra*	64
Stinging Nettle	122	Timothy Grass	138
Stinking Goose-foot	117	*Toxocampa*	53
„ Hellebore	81	Trailing Tormentil	96

	PAGE
Traveller's Joy	81
Trichiura	19
Trifolium	90
Trigonophora	45
Triphœna	38
Triphosa	77
Trochelium	10
Tropæolum	87
Tuberous Bitter Vetch	92
Tufted Vetch	92
Tussock Grass	139
Tussilago	106
Twayblade	136
Twig-rush	137
Typha	137

U.

Ulex	89
Ulmus	121
Unbranched Bur-reed	137
Uropteryx	54
Urtica	122

V.

Vaccinium	109
Valeria	45
Valeriana	103
Vanessa	4
Venilia	55
Venusia	60
Verbascum	113
Veronica	114
Vetch, Common	92
Viburnum	101
Vicia	92
Vinca	112
Vine	88
Viola	83
Viper's Bugloss	113
Vitis	88

W.

	PAGE
Wall Pennywort	99
Water Avens	95
,, Bedstraw	103
,, Cress	82
,, Dock	120
,, Figwort	114
Welted Thistle	107
Wheat	142
White Beak-rush	137
,, Beam	96
,, Campion	84
,, Dead-nettle	116
,, Goose-foot	117
,, Meadow Saxifrage	99
,, Mullein	113
,, Poplar	135
,, Willow	129
Whitethorn	97
Whortleberry	109
Wild Angelica	101
,, Carrot	101
,, Cherry	94
,, Mignonette	83
,, Mustard	82
,, Strawberry	95
,, Teasel	104
Winter Cress	82
Wood Anemone	81
,, Cow-wheat	114
,, Meadow-grass	141
,, Rush	137
,, Sage	116
,, Vetch	92
Woody Nightshade	113
Worm-seed Treacle Mustard	82
Wormwood, Common	105
Wych Elm	121

X.

Xanthia	41
Xylina	48

		PAGE
Xylocampa	. . .	48
Xylomiges	. . .	48
Xylophasia	. . .	29

		PAGE
Yellow Rattle	. .	114
„ Toad-flax	.	114
Yew	. . . .	136

Y.

Yarrow	. . .	105
Yellow Iris	. .	136
„ Meadow Rue	.	81

Z.

Zanclognatha	. .	53
Zeuzera	. . .	17
Zonosoma (Ephyra)	.	60
Zygæna	. . .	12

FOR NOTES, ETC.

FOR NOTES, ETC.

HERBERT W. MARSDEN,

Formerly of GLOUCESTER,

21, NEW BOND STREET, BATH.

Apparatus and Cabinets of all kinds for Entomological and other Natural History pursuits.

Specimens of British, European, and Exotic Lepidoptera, Coleoptera, &c.

Preserved Larvæ of British Lepidoptera in great variety.

Birds-Eggs, Birds-Skins, Shells, &c., &c.

New and Second-hand Books on Natural History Subjects.

Send addressed postal wrapper for Catalogue.

OVA, LARVÆ, and PUPÆ.

The best Stock in the Kingdom throughout the Season.

J. & W. DAVIS, Naturalists,

31 & 33a, HYTHE STREET, DARTFORD, KENT.

How to Rear Lepidoptera from Ova, Larvæ, & Pupæ, illust'd, 4d.

Hints on Egg Collecting and Nesting, 3½d.

Bird Stuffing for Amateurs, with 70 diagrams and illustrations. Best cheap work published, 7d.

Illustrated Catalogue of Entomological Apparatus, Naturalists' Requisites, Artificial Eyes, Taxidermists' Tools, Entomological Pins (silvered and black), Store Boxes, Cabinets, British Lepidoptera, Birds' Eggs and Skins, Stuffed Birds, Natural History Books, &c., &c. *The best List issued 2d. post free.*

Apparatus for Preserving Larvæ, 5/6

Larvæ Breeding Cages, 2/3; double, 4/6

Sweeping Net, 7/-; do., self-acting, 8/6

Beating Tray for Larvæ Beating, 10/6; or in case, 12/-

Oval Zinc Larvæ Boxes, 1/- and 1/6

Catalogue containing instructions for collecting and preserving all Natural History objects free for stamp.

JOHN EGGLESTON, Naturalist,

SUNDERLAND.

W. H. ALLEN & CO.'S
Works on Natural History.

Naturalist's Library. Edited by Sir WM. JARDINE, F.L.S., F.R.S. Containing numerous Portraits and Memoirs of Eminent Naturalists. Illustrated with 1,300 Coloured Plates. 42 vols. Fcp. 8vo, cloth, £9 9s. The Volumes may be had separately, 4s. 6d. each.

British Butterflies and Moths: an Illustrated Natural History of By EDWARD NEWMAN, F.Z.S. With over 800 Illustrations by George Willis and John Kirchner. Super-royal 8vo, cloth gilt, 25s.

The above work may also be had in Two Volumes, sold separately. Vol. I., Butterflies, 7s. 6d.; Vol. II., Moths, 20s.

Nature's Bye-Paths: a Series of Recreative Papers in Natural History. By J. E. TAYLOR, F.L.S., F.G.S. New Edition. Crn. 8vo., cloth, 3s. 6d.

The Aquarium: its Inhabitants, Structure, and Management. By J. E. TAYLOR, F.L.S., F.G.S., &c. With 238 Woodcuts. Second Edition. Crown 8vo, cloth, 3s. 6d.

The Collector's Handy-Book of Algæ, Diatoms, Desmids, FUNGI, LICHENS, MOSSES, &c. By JOHANN NAVE. Translated and Edited by the Rev. W. W. SPICER, M.A. Illustrated with 114 Woodcuts. Fcp. 8vo, cloth, 2s. 6d.

Notes on Collecting and Preserving Natural History OBJECTS. Edited by J. E. TAYLOR, F.L.S., F.G.S. New Edition. With numerous Illustrations. Crown 8vo, cloth, 3s. 6d.

The Preparation and Mounting of Microscopic Objects. By THOMAS DAVIES. New Edition, greatly enlarged and brought up to the present time by JOHN MATTHEWS, M.D., F.R.M.S. Fscp. 8vo, cloth, 2s. 6d.

Half-Hours with the Microscope. A Popular Guide to the Use of the Microscope as a means of Amusement and Instruction. By E. LANKESTER, M.D. With 250 Illustrations. Sixteenth Edition, enlarged Fcp. 8vo, plain, 2s. 6d.; coloured, 4s.

How to Choose a Microscope. By a Demonstrator. With 80 Illustrations. Demy 8vo, sewed, 1s.

LONDON: W. H. ALLEN & CO., 13, WATERLOO PLACE, S.W.

Established 1868.

W. K. MANN,

NATURALIST,

WELLINGTON TERRACE, CLIFTON, BRISTOL.

HAVING recently bought three large and valuable Collections of British Lepidoptera, I am able to offer many great rarities at moderate prices, and ordinary species exceedingly cheap.

A fine assortment of beautiful Varieties, selections of which may be had on approval.

A Special Price List of British Lepidoptera now ready (free).

A large stock of Pupæ during the winter, and Larvæ in the summer months.

SPECIAL OFFER.

SET OF INSECT COLLECTING APPARATUS,

including Net, Store Box, Setting Boards, Killing Box, Pins, &c., 6/6 free.

My New General Catalogue, 52 pp., now ready, free 1 stamp.

Collections of Lepidoptera, Eggs, Shells, &c., purchased for cash.

Old Collections of Stamps purchased or taken in exchange.

NATURALISTS' SUPPLY STORES

31, PARK STREET, WINDSOR.

Proprietor: E. EDMONDS.

Dealer in Entomological Apparatus of every description. 32 pp. Price List on application. Living *Ova, Larvæ* and *Pupæ* a SPECIALITY. (No larger stock in Europe.) Lists issued on 1st and 15th of each month: post free on receipt stamp, or 1/- per annum.

British, European & Exotic Species always on sale.

The New Improved Complete "*Larvæ Preserving Apparatus*," with full instructions for use, post free in Britain, 5/6.

31, PARK STREET, WINDSOR.

WATKINS & DONCASTER,

Naturalists,

36, STRAND, LONDON, W.C.

(Five doors from Charing Cross.)

Every description of Apparatus and Cabinets of the best make for Entomology and general Natural History, &c.

Wire or Cane Ring Net and Stick, 1s 8d., 2s., 2s. 3d. Umbrella Net (self-acting), 7s. 6d., Pocket Folding Net (wire or cane), 3s. 9d., 4s. 6d. Corked Pocket Boxes, 6d., 9d, 1s., 1s. 6d. Zinc Relaxing Boxes, 1s., 1s. 6d., 2s. Chip Boxes, nested, 4 dozen 8d.. Entomological Pins, mixed, 1s. per oz. Pocket Lantern, 2s. 6d., 5s., 10s. 6d. Sugaring Tin (with brush), 1s. 6d., 2s. Mite Destroyer, 1s. 6d. per lb. Best Killing Bottles, 1s. 6d. Store Boxes, 2s. 6d., 4s., 5s., and 6s. Setting Boards, from 6d., complete Set, 10s. 6d. Setting Houses, 9s. 6d., 11s. 6d., 14s. Larva Boxes, 9d., 1s., 1s. 6d. Breeding Cages, 2s. 6d., 4s., 5s., 7s. 6d.

A very fine and large Stock of British and Foreign Butterflies, Beetles, Birds' Eggs, &c.

As we receive parcels direct from our correspondents abroad, we are enabled to offer many good species of Lepidoptera from time to time.

Throughout the winter a large selection of Eggs, Chrysalises of British and Foreign Butterflies and Moths, including the gigantic Atlas and other Exotic Moths.

Collections of Natural-History objects, carefully named and arranged.

New and Second-hand Works on Entomology.

Label Lists of every description. The complete Label List of British Lepidoptera (Latin and English names), 1s. 6d. post free.

One each of all the British Butterflies in a case 25s.

A magnificent assortment of Preserved Caterpillars always in Stock.

Birds and animals stuffed and mounted in the best style by skilled workmen on the premises.

A full Catalogue sent, pp. 56, post free on application.

W. LONGLEY,

𝔈ntomological 𝔊abinet & 𝔄pparatus 𝔐aker,

WHOLESALE TO THE TRADE,

12, WHITE HART STREET;

CATHERINE STREET, STRAND, LONDON, W.C.

Nets, Breeding-cages and Apparatus of every description, Cabinets for Insects, Birds' Eggs, Minerals, Shells, Coins, etc., etc. Pocket-boxes, Store-boxes, and Book-boxes. Sheets of Cork any size.

NATURAL HISTORY AGENT AND BOOKSELLER.

Agent for *The Naturalists' Gazette*, post free, 1½d. monthly.

Setting boards for Insects, ½ in. 5d.; 1½ in. 8d.; 2 in. 10d.; 2½ in. 1s.; 3 in. 1s. 2d.
Cane, Net, and Stick, complete, 1s. 6d. Net Forceps, 2s. 6d.
Setting-houses, 9s. 6d., 11s. 6d. and 14s. Water Net, 2s. 6d.
Larvæ Breeding Cages, 2s. 6d., double, 5s.
    Do. do., with Water Cistern for keeping food fresh while rearing the Larvæ, 4s., double, 7s. 6d.
Zinc Larvæ Collecting Boxes, 9d., double, 1s., treble, 1s. 6d.
Sweeping Net, 8s. 6d. Nested Willow Chip Boxes, 4 doz., 8d.
Pocket Boxes (Deal), 6d., 9d., 1s. 1s. 6d.
Zinc Pocket and Relaxing Boxes, 9d., 1s. 1s. 6d., 2s.
Steel Forceps for removing Insects, post free for 1s. 6d.
Nickel Plated ditto   ,,   2s.
Cabinet Cork   7 × 3½   common, 1 dozen sheets, 1s.
    Do.   7 × 3½   best,   1   ,,   1s. 4d.
    Do,   11 × 4½   best,   1   ,,   2s. 8d.
Entomological Pins, ¼ oz. box, post free, 7½d.
Improved Japanned Collecting Box with straps for the shoulder, 9 in. × 7 in., post free, 5s. 6d.
Coleopterists' or Beetle Collecting Bottle, post free, 1s. 8d.
Large Sheets of Cork made to any size required @ 1s. 6d. per superficial foot.
Postal Boxes, 6d., 9d., 1s., 1s. 6d.
Cabinet, corked, glazed, and papered for Insects, 6 drawers, 17s. 6d.
    Do., 120 divisions for Birds' Eggs, 6 drawers, 16s. 6d.
    Do., of 6 drawers, for Minerals, 15s.
The New Larva Preserving Apparatus (Patented), 5s. 6d.
The New Birds' Egg Blowing Apparatus (Patented), 1s. 2d.
Cabinets, Store-boxes, Book-boxes, Setting-boards and houses, Pocket boxes and Nets of all sorts made to order on liberal terms to the trade by

W. LONGLEY,

12, WHITE HART St., CATHERINE St., STRAND, LONDON, W.C.

www.ingramcontent.com/pod-product-compliance
Lightning Source LLC
Chambersburg PA
CBHW020254170426
43202CB00008B/360